GLENCOE

PHYSICS

Principles and Problems

Study Guide

Student Edition

McGraw-Hill

New York, New York Columbus, Ohio Woodland Hills, California Peoria, Illinois

GLENCOE

PHYSICS

Principles and Problems

Student Edition

Teacher Wraparound Edition

Teacher Classroom Resources

- Transparency Package with Transparency Masters
- Laboratory Manual SE and TE
- Physics Lab and Pocket Lab Worksheets
- Study Guide SE and TE
- Chapter Assessment
- Tech Prep Applications
- Critical Thinking
- Reteaching
- Enrichment
- Physics Skills
- Advanced Concepts in Physics
- Supplemental Problems
- Problems and Solutions Manual
- Spanish Resources
- Lesson Plans with block scheduling

Technology

- Computer Test Bank (Win/Mac)
- MindJogger Videoquizzes
- Electronic Teacher Classroom Resources (ETCR)
- Website at *www.science.glencoe.com*
- Physics for the Computer Age CD-ROM (Win/Mac)

The Glencoe Science Professional Development Series

- Graphing Calculators in the Science Classroom
- Cooperative Learning in the Science Classroom
- Alternate Assessment in the Science Classroom
- Performance Assessment in the Science Classroom
- Lab and Safety Skills in the Science Classroom

Glencoe/McGraw-Hill

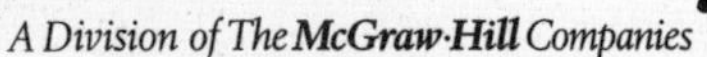
A Division of The McGraw-Hill Companies

Send all inquiries to:
Glencoe/McGraw-Hill
8787 Orion Place
Columbus, Ohio 43240

ISBN 0-02-825493-7
Printed in the United States of America.

11 12 13 14 045 08 07 06 05

Contents

To the Student

This *Study Guide* for ***Physics: Principles and Problems*** is a tool that will help you learn physics. It includes six pages of questions and exercises for each chapter of your textbook. The first *Study Guide* worksheet for each chapter is a review of vocabulary. The rest of the worksheets closely follow the organization of your textbook, providing review items for each textbook section and references to specific content.

You will find the directions in these worksheets simple and easy to follow. Diagrams are clear and useful. An assortment of question types are used: fill-in-the-blanks, matching, true/false, interpreting diagrams and data, multiple choice, short-answer questions, and so on.

Complete each *Study Guide* section after you have read the corresponding textbook section. When complete, the worksheets become an important tool that you can use to study for a test or just review the important ideas of the textbook.

Date ____________ Period ____________ Name ____________________

1 Study Guide

Use with Chapter 1.

What is physics?

Vocabulary Review

Write the term that correctly completes each statement. Use each term once.

physics | **quintessence** | **technology**
planets | **scientific method**

1. ______________ The study of matter and energy is _____.
2. ______________ A _____ is a systematic way to observe, experiment, and analyze the world.
3. ______________ The _____ are five bright objects that move relative to the constellations.
4. ______________ Greek philosophers thought that celestial bodies are made of the element _____.
5. ______________ The application of science to solve a practical problem is _____.

Section 1.1: Physics: The Search for Understanding

In your textbook, read about the work that physicists do.
Write the term that correctly completes each statement. Use each term once.

experiments | **observations** | **theories**
explanation | **phenomena** | **universe**
models | **predictions**

Physicists make **(1)** ______________, do **(2)** ______________, and create **(3)** ______________ or **(4)** ______________ to answer questions. Like all scientists, their goal is to obtain a compelling **(5)** ______________ that describes many different **(6)** ______________, makes **(7)** ______________, and leads to a better understanding of the **(8)** ______________.

Name ______________________________

1 Study Guide

In your textbook, read about early astronomy.
For each term on the left, write the letter of the matching item.

_____	**9.** wanderer	**a.**	Mars
_____	**10.** one of the elements that the ancient Greeks thought composed matter	**b.**	water
_____	**11.** the element that the ancient Greeks thought composed celestial objects	**c.**	planet
_____	**12.** the Roman name of the god of war	**d.**	quintessence
_____	**13.** the result of an element traveling in a straight line toward its own natural place	**e.**	motion

In your textbook, read about Galileo.
For each of the statements below, write true *or rewrite the italicized part to make the statement true.*

14. Galileo estimated the heights of the mountains of the moon by their *volumes.*

____

15. Galileo discovered that the planet *Mars* has four moons.

____

16. Galileo discovered that the planet *Venus* has phases.

____

17. Galileo argued that the planets circle *the sun.*

____

18. Galileo wrote most of his scientific works in his native *Latin,* so that they could be read by an educated person.

____

19. According to Galileo, the first step of studying a problem in an organized way is *observing.*

____

20. Galileo is considered to be the father of modern *space exploration.*

____

Name ______________________

1 Study Guide

In your textbook, read about scientific methods.
Label the following steps in the order in which scientists study problems.

_____ **21.** Draw a conclusion.

_____ **22.** Combine experimentation with careful measurements and analyses of results.

_____ **23.** Subject the conclusion to additional tests to determine if it is valid.

In your textbook, read about observations of Mars.
Answer the following question.

24. List four features attributed to Mars by astronomers and the general public that made it seem like Earth.

____

____

____

____

____

____

In your textbook, read about missions to Mars.
In the table below, enter each mission's launch date and check the appropriate accomplishment box.

Table 1

Mission	Launch Date	Result/Goal		
		Fly-by	Orbit	Land
Mariner 4 (U.S.)				
Mariner 6 (U.S.)				
Mariner 9 (U.S.)				
Mars 3 (USSR)				
Mars 5 (USSR)				
Viking 1 (U.S.)				
Viking 2 (U.S.)				
Global Surveyor (U.S.)				
Mars Pathfinder (U.S.)				
Mars Surveyor Orbiter (U.S.)				
Mars Surveyor Lander (U.S.)				

Name ______________________

1 Study Guide

In your textbook, read about the makeup of the Mars exploration teams.

Under each heading, list four careers that are represented on a Mars exploration team.

Scientific	Engineering	Other
______	______	______
______	______	______
______	______	______
______	______	______

In your textbook, read about the importance of physics.

Describe how physics is involved in each aspect of a Mars exploration mission.

25. Propulsion and guidance

____

26. Onboard energy systems

____

27. Communications

____

Name ______________________________

1 Study Guide

In your textbook, read about physics success stories.
For each term on the left, write the letter of the matching item.

_____ **28.** will allow performing functions on a submicroscopic level

_____ **29.** a result of research in heat transfer, thin films, plasma sources, vacuum technology , and new materials

_____ **30.** once thought worthless but now used in many endeavors including data storage, medicine, and telecommunications

_____ **31.** developed from wartime research in radar and miniature electronics

_____ **32.** aids in quality control

_____ **33.** developed from research into thin films, magnetic materials, and semiconductors

_____ **34.** helps reduce automotive costs

a. laser
b. microwave oven
c. personal computer
d. nanotechnology
e. lighter weight composite materials and polymers
f. energy efficient houses
g. computerized vision system

Date ______________ Period ______________ Name ______________________

2 Study Guide

Use with Chapter 2.

A Mathematical Toolkit

Vocabulary Review

Write the term that correctly completes each statement. Use each term once.

accuracy	**kilogram**	**precision**	**significant digits**
base units	**linear relationship**	**quadratic relationship**	**slope**
derived units	**meter**	**scientific notation**	***y*-intercept**
factor-label method	**metric system**	**second**	
inverse relationship	**parallax**	**SI**	

1. ______________ The apparent shift in the position of an object when it is viewed from various angles is _____.
2. ______________ The SI base unit of mass is the _____.
3. ______________ A way to convert a quantity in one unit to a quantity in another unit, using a relationship between the two units, is the _____.
4. ______________ The SI base unit of length is the _____.
5. ______________ Expressing numbers in exponential notation is _____.
6. ______________ On a graph, the rise divided by the run is the _____ of the line.
7. ______________ The system of measurement established by French scientists in 1795 is the _____.
8. ______________ A relationship that produces a graph with a straight line is a(n) _____.
9. ______________ The degree of exactness of a measurement is _____.
10. ______________ The units of quantities measured against standards are the _____.
11. ______________ A relationship in which one variable varies with the square of the other variable is a(n) _____.
12. ______________ On a graph, the point where the line crosses the y-axis is the _____.
13. ______________ The Système Internationale d'Unités, _____, is a system of measurement established by an international committee.
14. ______________ A relationship in which one variable varies with the inverse of the other is a(n) _____.
15. ______________ Units that are combinations of base units are _____.
16. ______________ In any measurement, the digits that are valid are _____.
17. ______________ The SI base unit of time is the _____.
18. ______________ The extent to which a measured value agrees with the standard value of the quantity is _____.

Name ______________________________

2 Study Guide

Section 2.1: The Measures of Science

In your textbook, read about SI.

For each base quantity on the left, write the matching base unit and the matching symbol.

				Base Unit	Symbol
1.	________	_____	length	ampere	cd
2.	________	_____	mass	candela	kg
3.	________	_____	time	kelvin	m
4.	________	_____	temperature	kilogram	mol
5.	________	_____	amount of a substance	meter	s
6.	________	_____	electric current	mole	A
7.	________	_____	luminous intensity	second	K

Answer the following questions, using complete sentences.

8. What is the current definition of a meter?

____

____

____

____

____

9. How is a second currently defined?

____

____

____

____

____

10. What is the current definition of a kilogram?

____

____

____

____

____

Name ______________________

2 Study Guide

In your textbook, read about SI prefixes.
Complete the table.

Table 1			
Prefix	**Symbol**	**Multiplier**	**Scientific Notation**
11.		1/1 000 000 000 000 000	
12. pico			
13.			10^{-9}
14. micro			
15.		1/1000	
16.		1000	
17. mega			
18.			10^{9}
19.	T		

Write the term that correctly completes each statement.

20. In scientific notation, the numerical part of a quantity is written as ______________________.

21. In 10^n, n is an ______________________.

22. In changing the numerical part of a measurement to scientific notation, the number of places you move the decimal point to the left is expressed as a ______________________ exponent.

23. In changing the numerical part of a measurement to scientific notation, the number of places you move the decimal point to the right is expressed as a ______________________ exponent.

Name ______________________

2 Study Guide

Section 2.2: Measurement Uncertainties

In your textbook, read about accuracy and precision.

Refer to the diagram below. Circle the letter of the choice that best completes each statement.

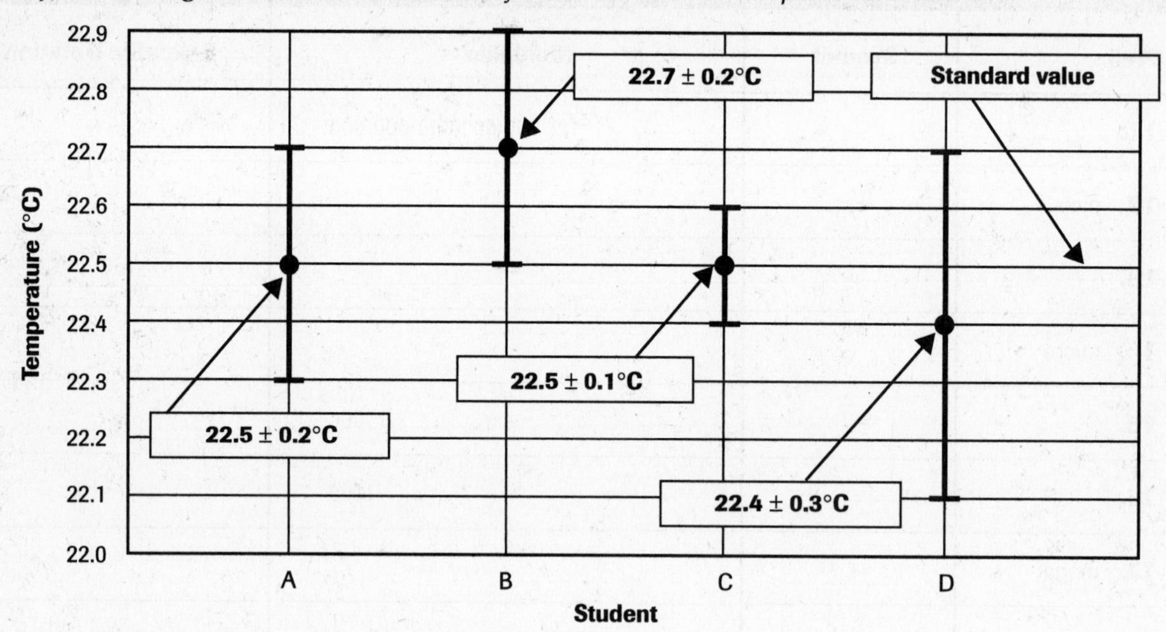

1. Measurements that have the same precision are those of students _____.

a. A and B **b.** A and C **c.** A and D **d.** B and D

2. Measurements that have the least precision are those of student _____.

a. A **b.** B **c.** C **d.** D

3. Measurements that have the greatest precision are those of student _____.

a. A **b.** B **c.** C **d.** D

4. Measurements that have the same accuracy are those of students _____.

a. A and B **b.** A and C **c.** A and D **d.** B and D

5. Measurements that have the least accuracy are those of student _____.

a. A **b.** B **c.** C **d.** D

6. Measurements that have the greatest accuracy are those of students _____.

a. A and B **b.** A and C **c.** A and D **d.** B and D

7. The most precise and accurate measurements are those of student _____.

a. A **b.** B **c.** C **d.** D

8. To assure the greatest precision and accuracy in reading an instrument such as a thermometer, you should read the instrument _____.

a. below eye level **b.** above eye level **c.** at eye level

Name ____________________

2 Study Guide

In your textbook, read about significant digits.

Write the letter of each rule that is used to determine the correct number of significant digits in each measurement below. Then write the measurement in scientific notation.

Measurement	Rules Used	Scientific Notation
9. 321 m	________	____________
10. 1201 s	________	____________
11. 1.0 m/s	________	____________
12. 0.01 K	________	____________
13. 70.010 Mg	________	____________
14. 53 000 000 cm	________	____________
15. 0.200 A	________	____________
16. 40 000 L	________	____________
17. 0.680 kg	________	____________

Rules

a. Nonzero digits are always significant.

b. All final zeros after the decimal point are significant.

c. Zeros between two other significant digits are always significant.

d. Zeros used solely as placeholders are not significant.

In your textbook, read about arithmetic with significant digits.

Explain how the number of significant digits was determined for the answer to each problem.

18.
$$\begin{array}{r} 21.65 \text{ mm} \\ 14.00 \text{ mm} \\ +\ 1.2 \text{ mm} \\ \hline 37.9 \text{ mm} \end{array}$$

____

____

____

19.
$$\begin{array}{r} 0.62 \text{ m} \\ \times\ 3.57 \text{ m} \\ \hline 2.2 \text{ m}^2 \end{array}$$

____

____

____

20. $\dfrac{5013 \text{ m}}{2.6 \text{ s}} = 1900 \text{ m/s}$

____

____

____

Name ______________________

2 Study Guide

Section 2.3: Visualizing Data

In your textbook, read about graphing data.

Read the paragraph below. Refer to the table to answer the questions that follow.

A student cuts an aluminum rod into four random pieces. She then measures the volume and mass of each piece. The data she records is shown. To determine how the mass of each piece varies with its volume, she constructs a graph.

Table 1	
Volume (mL)	**Mass (g)**
9.2	24.93
18.3	49.15
34.5	93.19
48.2	131.92

1. What is the range of the data of the independent variable?

____

2. What would be a convenient range for the *x*-axis?

____

3. What is the range of the data of the dependent variable?

____

4. What would be a convenient range for the *y*-axis?

____

In your textbook, read about linear and nonlinear relationships.

For each term on the left, write the letter of the matching item.

_____ **5.** the equation of a linear relationship

_____ **6.** the shape of the graph of a linear relationship

_____ **7.** the equation of an inverse relationship

_____ **8.** the shape of the graph of an inverse relationship

_____ **9.** the equation of a quadratic relationship

_____ **10.** the shape of the graph of a quadratic relationship

a. hyperbola

b. parabola

c. straight line

d. $y = mx + b$

e. $y = ax^2 + bx + c$

f. $y = \frac{a}{x}$

Date ______________ Period ______________ Name ______________________________

3 Study Guide

Use with Chapter 3.

Describing Motion

Vocabulary Review

Write the term that correctly completes each statement. Use each term once.

average acceleration	motion diagram	position vector
average speed	operational definition	problem-solving strategy
average velocity	origin	scalar quantity
coordinate system	particle model	time interval
displacement	physical model	vector quantity
distance	pictorial model	

1. ______________________ A series of images of an object that records the object's position after equal time intervals is a(n) _____.
2. ______________________ A quantity that has magnitude and direction is a(n) _____.
3. ______________________ The _____ replaces an object by a single point.
4. ______________________ The zero point of a variable and the direction in which the values of the variable increase is indicated in a(n) _____.
5. ______________________ An arrow that is proportional to the distance an object is from the origin and points in the direction of the object is a(n) _____.
6. ______________________ The magnitude of the average-velocity vector is _____.
7. ______________________ A step-wise procedure for solving problems is a(n) _____.
8. ______________________ A change in time is a(n) _____.
9. ______________________ A quantity defined in terms of the procedure used to identify it is a(n)_____.
10. ______________________ The change in position of an object is its _____.
11. ______________________ The point in a coordinate system at which the variables have the value of zero is the _____.
12. ______________________ The length of the displacement vector represents the _____ between the two positions.
13. ______________________ A model that uses symbols to represent variables is a(n) _____.
14. ______________________ The ratio of the displacement to the time interval in which the displacement takes place is the _____.
15. ______________________ A motion diagram is an example of a(n) _____.
16. ______________________ The _____ is the ratio of the change in velocity to the time interval in which the change took place.
17. ______________________ A quantity that has only magnitude is a(n) _____.

Name ____________________

3 Study Guide

Section 3.1: Picturing Motion

In your textbook, read about motion diagrams.

Refer to the diagrams below, showing frames from a camcorder at 2-s intervals. Circle the letter of the choice that best completes each statement.

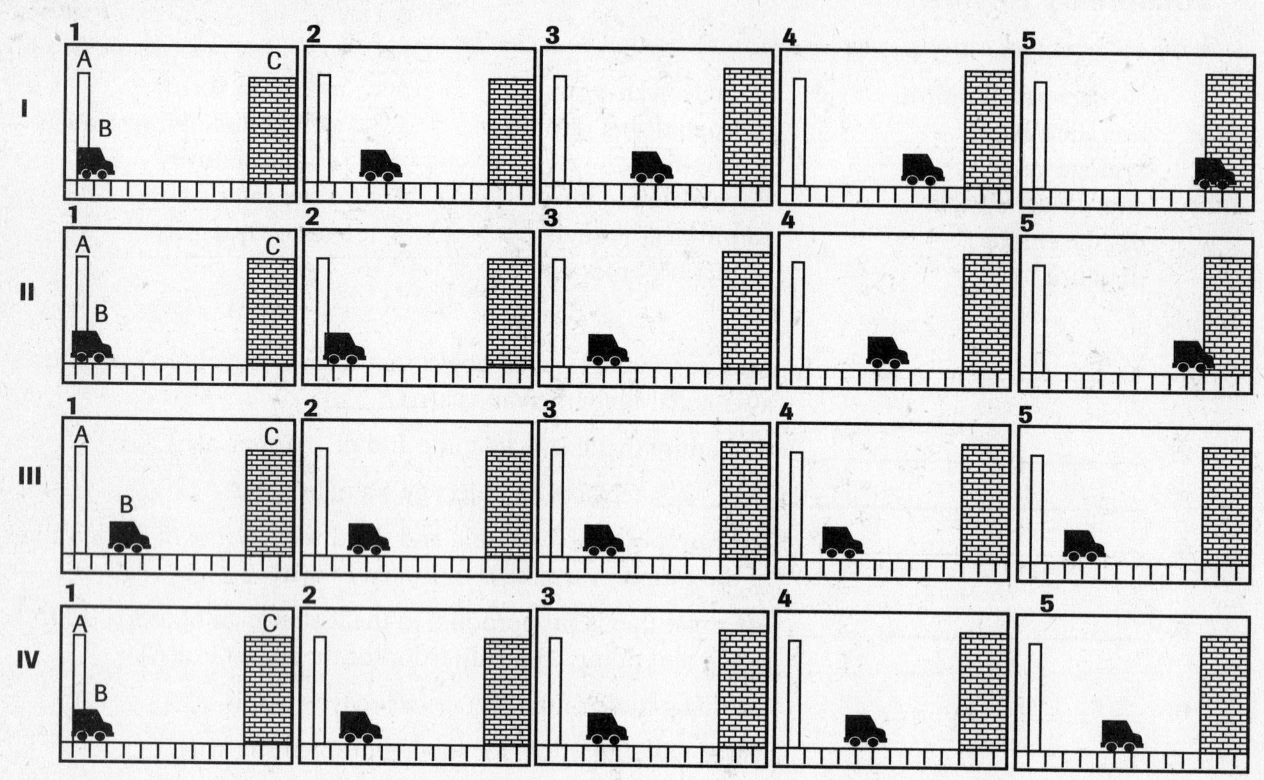

1. In set I, the object that is moving is _____.

a. A **b.** B **c.** C

2. Set II shows that object B is _____.

a. at rest

b. increasing its speed

c. slowing down

d. traveling at constant speed

3. Set _____ shows object B is slowing down.

a. I **b.** II **c.** III **d.** IV

4. Set _____ shows object B at rest.

a. I **b.** II **c.** III **d.** IV

5. Set _____ shows object B traveling at a constant speed.

a. I **b.** II **c.** III **d.** IV

Name ____________________

3 Study Guide

Section 3.2: Where and When?

In your textbook, read about coordinate systems.

Refer to the diagrams below, showing the location of an object represented by a circle. Complete the table by writing the position vector of each object.

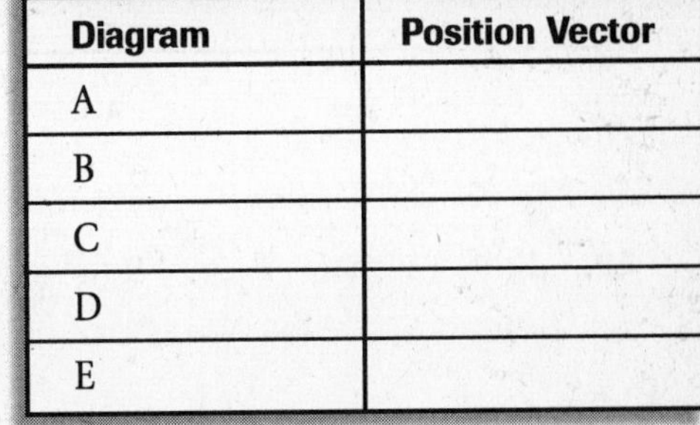

Table 1	
Diagram	**Position Vector**
A	
B	
C	
D	
E	

A

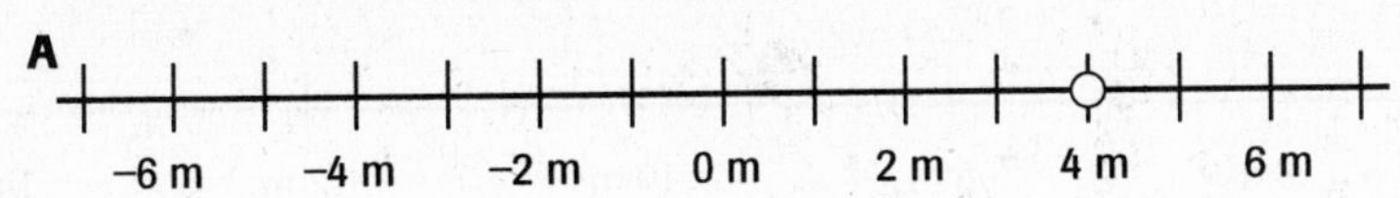

B

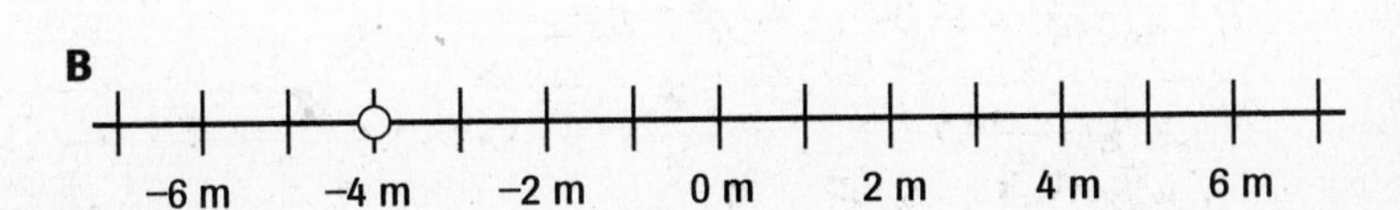

C

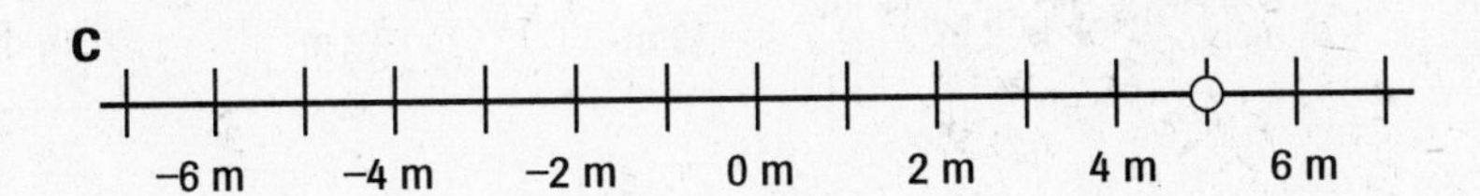

D

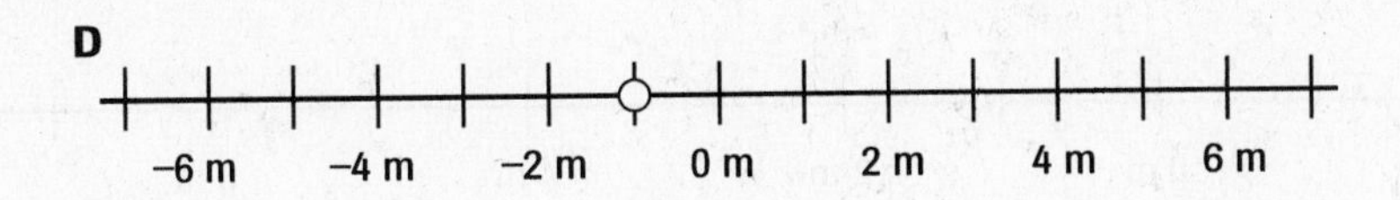

E

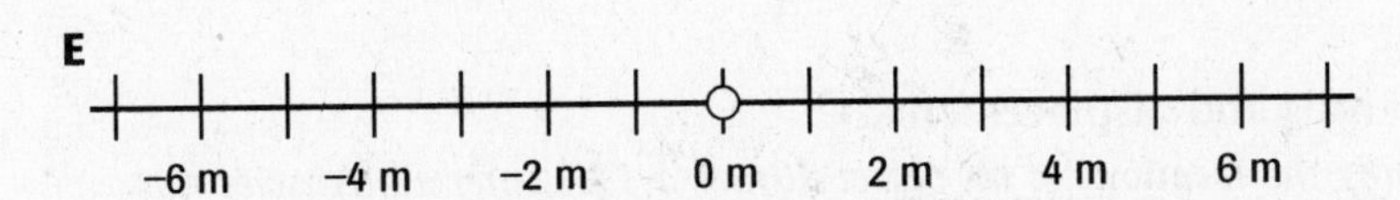

In your textbook, read about vectors, scalars, time intervals, and displacements.

For each term on the left, write the letter of the matching term.

_____ **1.** symbol that represents time interval

_____ **2.** Greek letter delta used to mean change

_____ **3.** definition of time interval

_____ **4.** one way of representing the vector quantity acceleration

_____ **5.** symbol that represents position

_____ **6.** magnitude of the displacement vector

_____ **7.** definition of displacement

_____ **8.** one way of representing the vector quantity velocity

_____ **9.** symbol that represents displacement

_____ **10.** symbol that represents the scalar quantity mass

a. distance
b. m
c. Δt
d. $\vec{a}$
e. Δd
f. Δ
g. v
h. d
i. $t_1 - t_0$
j. $d_1 - d_0$

Name ____________________

3 Study Guide

In your textbook, read about time intervals and displacements.

In the space provided, draw and label the vectors $\mathbf{d}_0$ *and* $\mathbf{d}_1$ *for the given information. Then draw and label the vector* $\Delta\mathbf{d}$ *and determine its magnitude.*

11. $d_0 = +4$ m $\quad d_1 = +10$ m

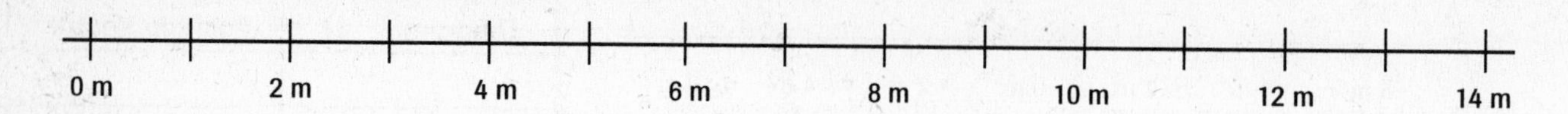

12. $d_0 = +2$ m $\quad d_1 = +9$ m

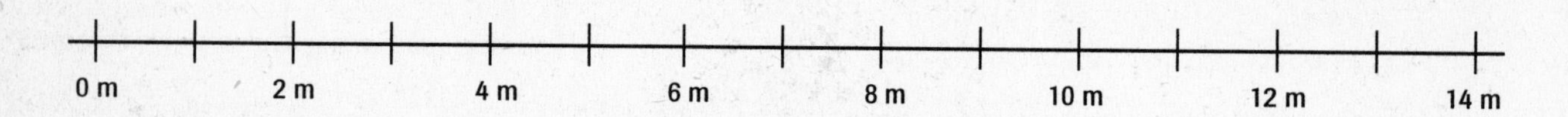

13. $d_0 = -4$ m $\quad d_1 = -6$ m

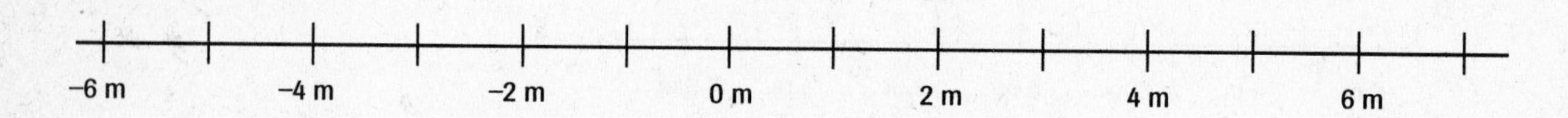

In your textbook, read about time intervals and displacements.

Refer to the motion diagram below, showing the locations of an object during a 12-s interval. Answer the following questions.

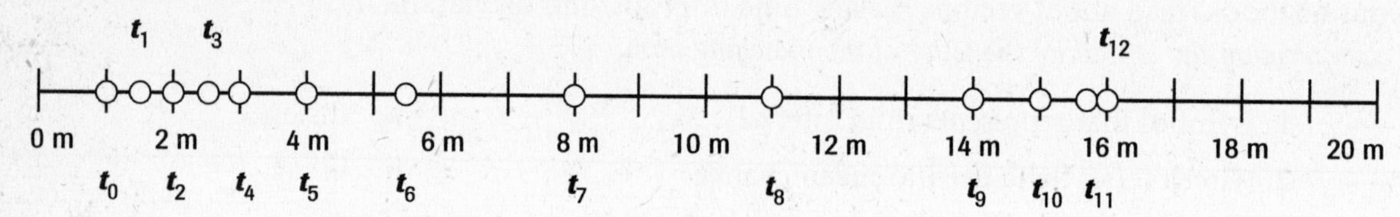

14. What is the magnitude of the position vector at $t = 0$ s?

____

15. At what time does the object have a position of +8.0 m?

____

16. What is displacement of the object in the time interval $t_6 - t_5$?

____

17. In what time intervals does the object have a constant speed?

____

18. What is happening to the motion of the object in the time interval $t_{12} - t_9$?

____

19. What distance does the object travel in the time interval $t_{12} - t_0$?

____

Section 3.3: Velocity and Acceleration

In your textbook, read about velocity.

For each of the statements below, write true *or rewrite the italicized part to make the statement true.*

1. ____________________ The ratio $\Delta d/\Delta t$ is called the *average velocity.*

2. ____________________ The symbol $\bar{v}$ represents the *average acceleration.*

3. ____________________ The magnitude of the average-velocity vector is the quantity average *distance traveled.*

4. ____________________ Automobile speeds are usually measured in miles per hour (mph) or *kilometers per hour (km/h).*

5. ____________________ The equation $d_0 + \bar{v}\,\Delta t$ represents the *position* of an object after the time interval Δt.

6. ____________________ The quantity $\bar{v}\,\Delta t$ represents the change in *speed* of an object during the time interval Δt.

In your textbook, read about interpreting velocity vectors and acceleration.

Refer to the motion diagram below, showing the position at 1-s intervals of a car traveling to the right along the positive x–axis. In the space provided, draw the displacement, velocity, and acceleration vectors for each time interval. Answer the following questions.

• • • • • • • • • •

Displacement

Velocity

Acceleration

7. What is happening to the motion of the car during the time intervals in which both the velocity and the acceleration vectors are in the positive direction?

____

8. What is happening to the motion of the car during the time intervals in which the velocity vectors are positive and the acceleration vectors are negative?

____

Name ______________________

3 Study Guide

In your textbook, read about problem-solving strategies.

Read the following problem and refer to the diagram below. Circle the letter of the choice that best completes the statement or answers the question.

School Zone

A car traveling at a constant 7 m/s takes 5 s to pass through a school zone and then accelerates at 2 m/s^2 for 3 s. How far does the car travel in the 8 s?

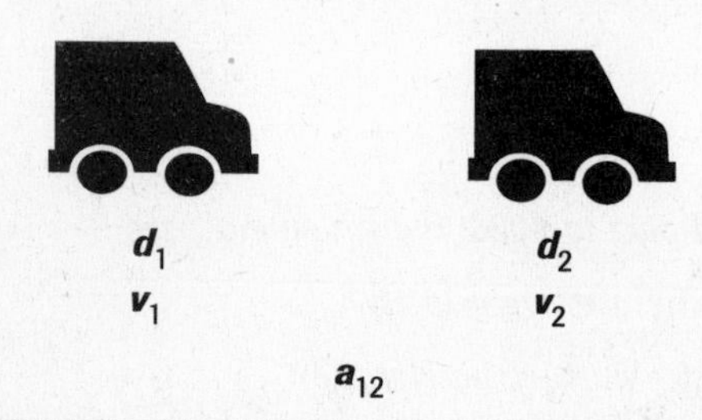

d_1 v_1 d_2 v_2 d_3 v_3

a_{12} a_{23} $+x$

1. The diagram above represents a(n) _____ of the problem.
   - a. physical model
   - b. operational definition
   - c. pictorial model
   - d. particle model
2. A likely choice for the origin of the coordinate system would be at a point _____.
   - a. where the car enters the school zone
   - b. halfway through the school zone
   - c. at the beginning of the school zone
   - d. at the end of the school zone
3. Which of the following motion diagrams represents the problem?
   - a. •••• • ••••
   - b. • • • ••••••
   - c. •••••••••
   - d. ••••• • • •
4. The problem can be thought of as a two-step problem because the car has _____.
   - a. one displacement
   - b. constant velocity and an acceleration
   - c. equal time intervals
   - d. unequal time intervals
5. If d_1 is the position of the car at the beginning of the problem and d_3 is the position of the car at the end of the problem, a good choice of d_2 is the position of the car _____.
   - a. at the beginning of the school zone
   - b. halfway through the school zone
   - c. at the point where the car changes speeds
   - d. at the point where the car stops
6. The acceleration a_{12}, between d_1 and d_2, is _____.
   - a. increasing
   - b. decreasing
   - c. 0 m/s^2
   - d. +2.0 m/s^2
7. The acceleration a_{23}, between d_2 and d_3, is _____.
   - a. increasing
   - b. decreasing
   - c. 0 m/s^2
   - d. +2.0 m/s^2
8. The vector representing acceleration a_{23} _____.
   - a. points to the left
   - b. points to the right
   - c. is equal to zero
   - d. is shorter than vector a_{12}

Date ______________ Period ______________ Name ______________________________

4 Study Guide

Use with Chapter 4.

Vector Addition

Vocabulary Review

Write the term that correctly completes each statement. Use each term once.

algebraic representation	**magnitude**	**vector**
components	**origin**	**vector decomposition**
direction	**Pythagorean relationship**	**velocity**
displacement	**resultant vector**	
graphical representation	**scalar quantity**	

1. ______________________ An arrow or an arrow-tipped line segment is called a(n) _____ of a vector.

2. ______________________ A(n) _____ of a vector is represented by an italicized letter in boldface type.

3. ______________________ A(n) _____ is a vector that is equal to the sum of two or more vectors.

4. ______________________ The _____ states that the sum of the squares of the two sides of a right triangle is equal to the square of the hypotenuse.

5. ______________________ The _____ of a vector is defined as the angle that the vector makes with the *x*-axis, measured counterclockwise, in a rectangular coordinate system.

6. ______________________ The process of breaking a vector into its components is sometimes called _____.

7. ______________________ Two vectors projected on the axes of a coordinate system are called the _____ of the resultant vector.

8. ______________________ A quantity that has both magnitude and direction is a(n) _____.

9. ______________________ The _____ of a vector is always a positive quantity.

10. ______________________ Distance is the magnitude of a vector quantity called _____.

11. ______________________ A quantity that has only magnitude is a(n) _____.

12. ______________________ The magnitude of an object's _____ is its speed.

13. ______________________ The center of a coordinate system is called the _____.

Name ______________________________

4 Study Guide

Section 4.1: Properties of Vectors

In your textbook, read about representing vector quantities.

For the diagrams below, use the scale to determine the magnitude of each displacement vector. Determine the direction of each displacement using compass points of north, south, east, and west, or the intermediate points such as northeast and southwest. Write the value of the displacement in the space to the left.

1. ______________________
2. ______________________
3. ______________________
4. ______________________
5. ______________________
6. ______________________
7. ______________________
8. ______________________
9. ______________________
10. ______________________

1. = 1 km

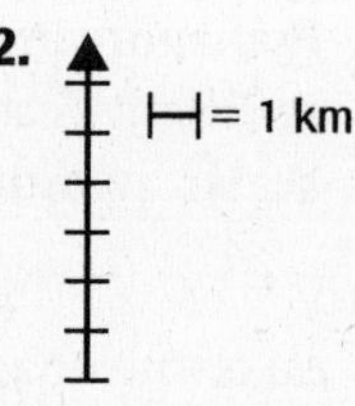

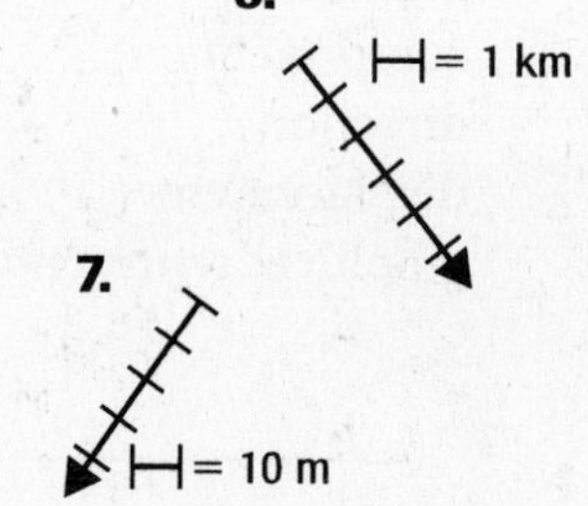

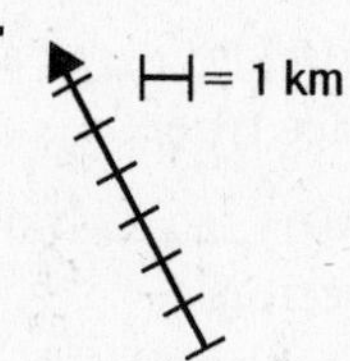

8. = 2 km

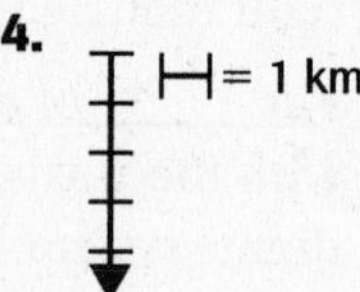

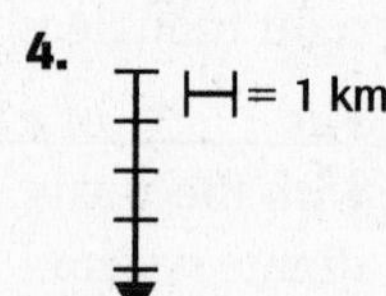

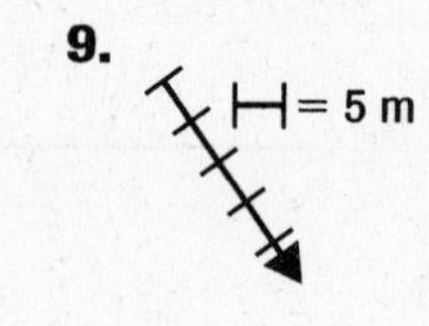

5. = 1 km

10. = 100 m

Read the following statement. Circle the letter of the choice that is the best response.

The displacement of a student is 2.0 km north.

11. Which is the correct algebraic representation of the displacement?

a. d = 2.0 km north

b. *d* = 2.0 km north

c. **d** = 2.0 km north

d. ***d*** = 2.0 km north

12. Which is the correct algebraic representation of the magnitude of the displacement?

a. d = 2.0 km

b. *d* = 2.0 km

c. **d** = 2.0 km

d. ***d*** = 2.0 km

Name ______________________

4 Study Guide

In your textbook, read about representing vector quantities.

Each set of displacement vectors below is drawn to the same scale. For each situation, write its letter next to the correct number of the expression.

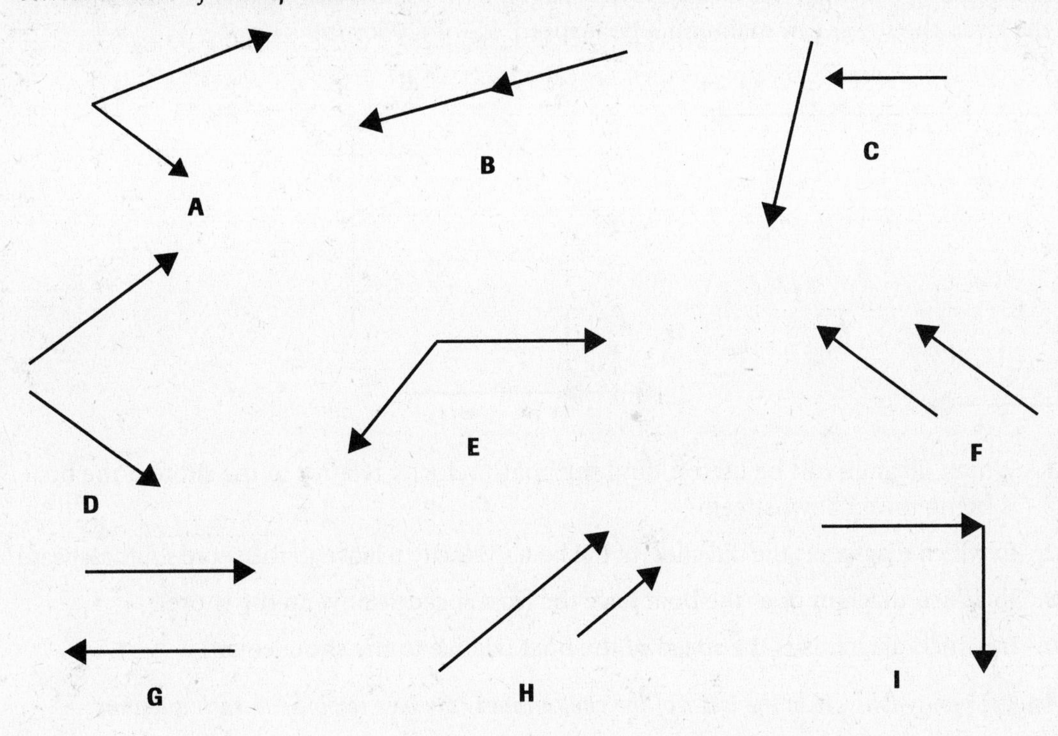

13. ______________ Vectors in set have same direction.

14. ______________ Vectors in set have same magnitude.

15. ______________ Vectors in set are equal.

In your textbook, read about the graphic addition of vectors.

For each of the statements below, write true *or rewrite the italicized part to make the statement true.*

16. ______________ The *length* of the arrow representing a vector should be proportional to the magnitude of the quantity being represented.

17. ______________ A *ruler* is used to measure the angle that establishes the direction of the vector.

18. ______________ When adding two vectors, the first vector is drawn and the *tip* of the second vector is placed at the tip of the first vector.

19. ______________ When adding vectors, the resultant vector is drawn from the *tail* of the first vector to the tip of the last vector.

20. ______________ The resultant vector is *dependent* on the order in which the vectors are added.

4 Study Guide

In your textbook, read about relative velocities.

Read the following paragraph. Write the letter of the diagram that is the best response to each question.

A boat is placed in a river, which has a current velocity, v_c, of 3.0 km/h east, relative to the shore. Relative to the river, the rower can maintain a boat speed, v_b, of 6.0 km/h.

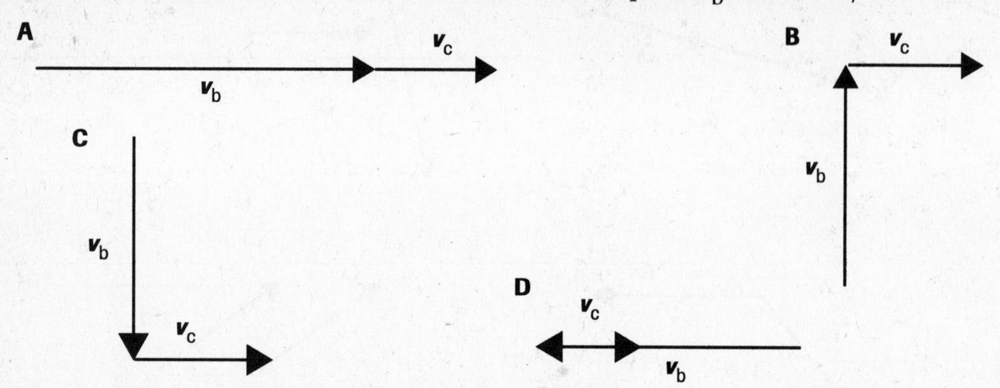

_____ **21.** Which diagram can be used to find the boat's velocity relative to the shore if the boat is being rowed downstream?

_____ **22.** In which diagram is the direction of the boat's velocity relative to the shore southeastward?

_____ **23.** In which diagram does the boat have the least speed relative to the shore?

_____ **24.** In which diagrams is the speed of the boat relative to the shore equal?

Read the following paragraph. Circle the letter of the choice that is the best response to each question.

In questions 21–24, the resultant vector represents the velocity of the boat, v_s, relative to the shore. Algebraically,

$$v_s = v_b + v_c.$$

If a rower wants to get to a certain point by a certain time, he or she knows what v_s is. What the rower must determine is v_b, that is, how fast and in what direction he or she must row relative to the river. Solving the above equation for v_b gives

$$v_b = v_s - v_c.$$

25. The equation for v_b can be written as which of the following equations?

a. $v_b = v_c - v_s$

b. $v_b = v_s + (-v_c)$

c. $v_b = -(v_s + v_c)$

d. $v_b = -(v_s - v_c)$

26. Which diagram illustrates the velocity at which a boat must be rowed relative to the river if the rower wants to maintain a heading of 45.0° northeast relative to the shore? The river has a current velocity of 3.0 m/s east relative to the shore.

a.

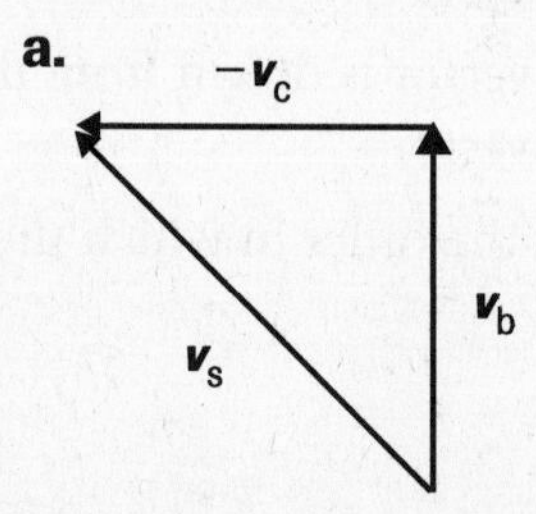

b.

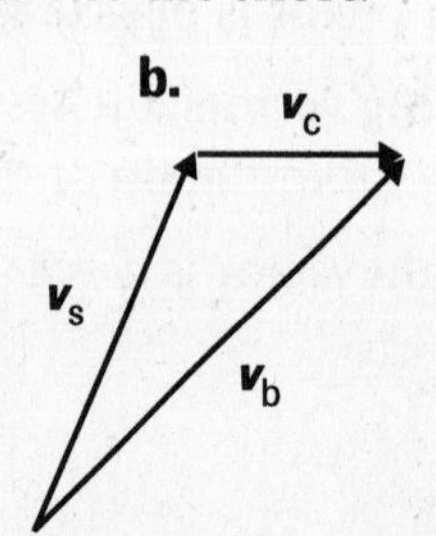

c.

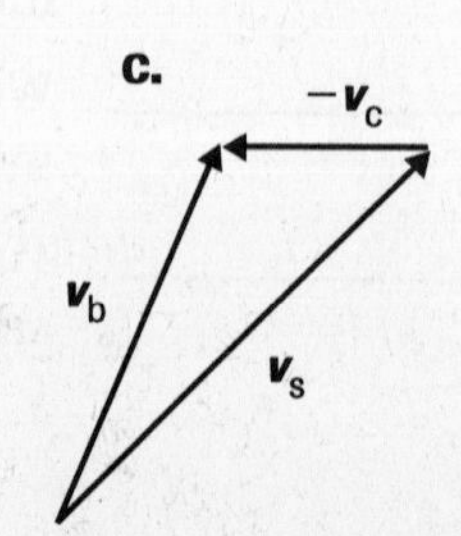

d.

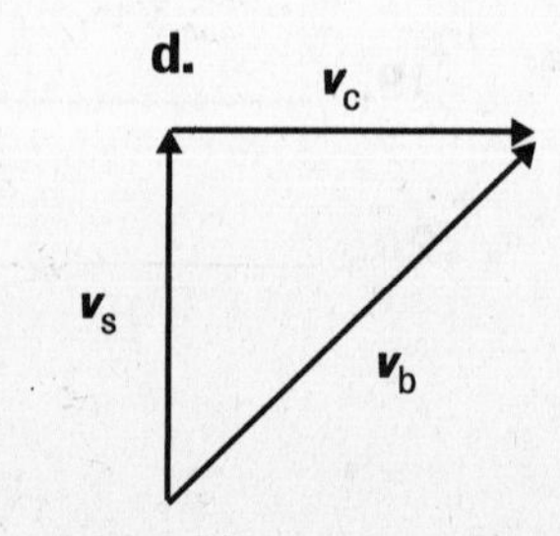

Name ______________________

4 Study Guide

Section 4.2: Components of Vectors

In your textbook, read about coordinate systems.

Refer to the diagram below. Circle the letter of the choice that best completes each statement.

1. In the coordinate system shown, the positive y-axis is located _____ counterclockwise from the positive x-axis.

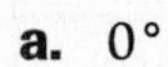

a. 0° **c.** 180°

b. 90° **d.** 270°

2. The y-axis crosses the x-axis at the _____.

a. angle θ **c.** quadrant

b. origin **d.** vertex

3. Vectors **E** and **F** lie in quadrant _____.

a. I **c.** III

b. II **d.** IV

4. Vector _____ has a positive horizontal component and a positive vertical component.

a. *A* **c.** *D*

b. *B* **d.** *E*

5. The sine of an angle is defined as

$$\sin\theta = \frac{\text{opposite side}}{\text{hypotenuse}}.$$

The sine of the angle that vector **B** makes with the x-axis is _____.

a. $\sin\theta = \frac{B_x}{B}$ **b.** $\sin\theta = \frac{B}{B_x}$ **c.** $\sin\theta = \frac{B_x}{B_y}$ **d.** $\sin\theta = \frac{B_y}{B}$

6. The cosine of an angle is defined as

$$\cos\theta = \frac{\text{adjacent side}}{\text{hypotenuse}}.$$

The cosine of the angle that vector **D** makes with the x-axis is _____.

a. $\cos\theta = \frac{D_x}{D}$ **b.** $\cos\theta = \frac{D}{D_x}$ **c.** $\cos\theta = \frac{D_x}{D_y}$ **d.** $\cos\theta = \frac{D_y}{D}$

7. Vector _____ has no horizontal component.

a. *A* **b.** *B* **c.** *C* **d.** *G*

8. The horizontal component of vector *F* has a value of _____.

a. F **b.** $F\sin\theta$ **c.** $F\cos\theta$ **d.** $\frac{F}{\cos\theta}$

9. The vertical component of vector *F* has a value of _____.

a. F **b.** $F\sin\theta$ **c.** $F\cos\theta$ **d.** $\frac{F}{\cos\theta}$

10. Both components of vector _____ are negative.

a. *B* **b.** *D* **c.** *E* **d.** *G*

Name ______________________

4 Study Guide

In your textbook, read about addition of vectors using components.

The diagram shows the displacement of a hiker. The data table shows the magnitudes of the vertical and horizontal components of each displacement.

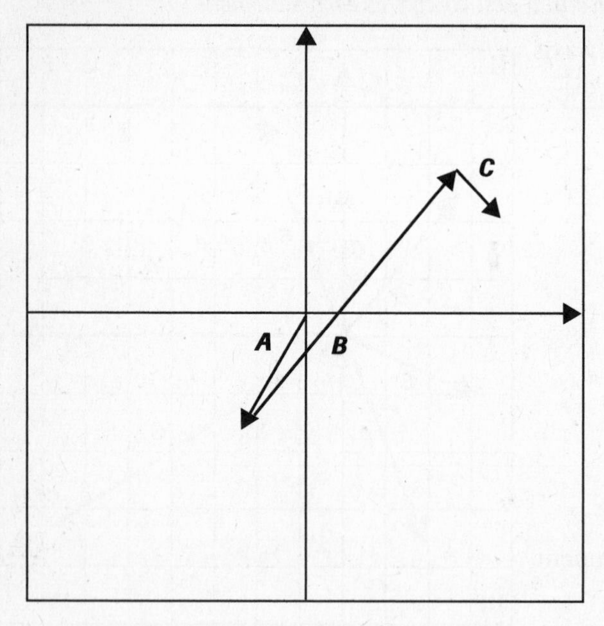

Table 1

Component	Magnitude (km)
A_x	1.4
A_y	3.8
B_x	4.0
B_y	6.9
C_x	1.5
C_y	1.3

Circle the letter of the choice that is the best response.

11. What is the value of R_x, which represents the sum of the horizontal components of the displacements?

a. $R_x = 1.4\ \text{km} + 4.0\ \text{km} + 1.5\ \text{km} = 6.9\ \text{km}$

b. $R_x = (-1.4\ \text{km}) + (-4.0\ \text{km}) + (-1.5\ \text{km}) = -6.9\ \text{km}$

c. $R_x = (-1.4\ \text{km}) + 4.0\ \text{km} + 1.5\ \text{km} = 4.1\ \text{km}$

d. $R_x = 1.4\ \text{km} + (-4.0\ \text{km}) + (-1.5\ \text{km}) = -4.1\ \text{km}$

12. What is the value of R_y, which represents the sum of the vertical components of the displacements?

a. $R_y = 3.8\ \text{km} + 6.9\ \text{km} + 1.3\ \text{km} = 12.0\ \text{km}$

b. $R_y = (-3.8\ \text{km}) + (-6.9\ \text{km}) + (-1.3\ \text{km}) = -12.0\ \text{km}$

c. $R_y = (-3.8\ \text{km}) + 6.9\ \text{km} + (-1.3\ \text{km}) = 1.8\ \text{km}$

d. $R_y = 3.8\ \text{km} + (-6.9\ \text{km}) + (-1.3\ \text{km}) = -4.4\ \text{km}$

13. What is the magnitude of the resultant R?

a. $R < 0$ km **b.** $R = 0$ km **c.** $R > 0$ km

14. What is the magnitude and direction of the displacement vector Y that will return the hiker to the starting point?

a. $Y = 0$ km **b.** $Y = R$ **c.** $Y = -R$ **d.** $Y = -2R$

Date ______________ Period ______________ Name ______________________________

5 Study Guide

Use with Chapter 5.

A Mathematical Model of Motion

Vocabulary Review

Write the term that correctly completes each statement. Use each term once.

acceleration due to gravity	**instantaneous velocity**	**slope**
constant acceleration	**position-time graph**	**uniform motion**
constant velocity	**rise**	**velocity-time graph**
instantaneous acceleration	**run**	

1. ______________________ A graph that shows how position depends on time is a(n) _____.

2. ______________________ The vertical separation of any two points on a curve is the _____.

3. ______________________ On a position-time graph, the slope of a tangent to the curve at a specific time is the _____.

4. ______________________ In _____ equal displacements occur during successive equal time intervals.

5. ______________________ An object that has the same average velocity for all time intervals is moving at _____.

6. ______________________ The constant acceleration that acts on falling bodies is the _____.

7. ______________________ The ratio of the rise to run is the _____ of a curve.

8. ______________________ A graph that shows how velocity depends on time is a(n) _____.

9. ______________________ Motion that can be described by a constant slope on a velocity-time graph is _____.

10. ______________________ The horizontal separation of any two points on a curve is the _____.

11. ______________________ On a velocity-time graph, the slope of a tangent to the curve at a specific time is the _____.

Name ______________________

5 Study Guide

Section 5.1: Graphing Motion in One Dimension

In your textbook, read about position-time graphs.

Answer the following questions, using complete sentences.

1. How are the position and time of a moving object related on a position-time graph?

   ____

2. If you decide that the duration of an instant on a position-time graph is a finite period of time, where is the object during that period?

   ____

3. What does your conclusion in problem 2 indicate about the motion of the object during that period?

   ____

4. If the object is moving, what can you conclude about the duration of an instant?

   ____

5. What is the duration of an instant?

   ____

In your textbook, read about using graphs to determine position and time.

The position-time graph below shows the position of a teacher at various times as she walks across the front of the room. The position 0.0 m represents the left side of the room and movement to the right is positive. Circle the letter of the choice that best completes each statement.

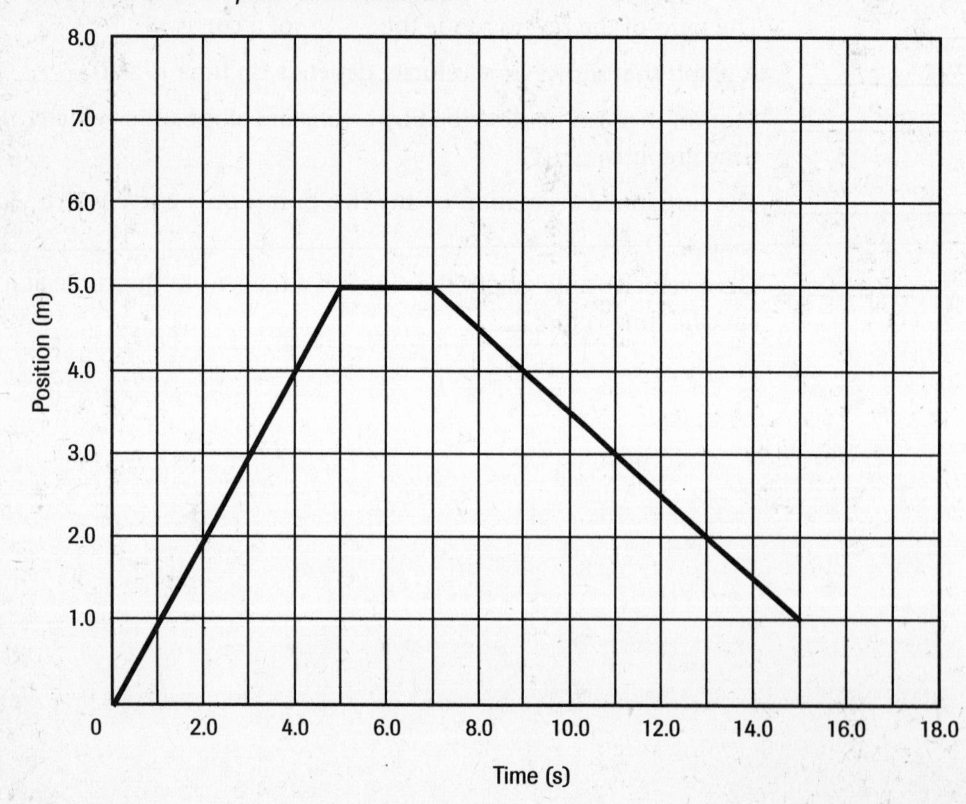

Name ______________________________

5 Study Guide

6. The teacher was walking to the right during the time interval _____.
   a. 1.0–2.0 s **b.** 6.0–7.0 s **c.** 10.0–12.0 s **d.** 7.0–12.0 s
7. The teacher's displacement for the time interval 0.0–3.0 s is _____.
   a. –3.0 m **b.** 0.0 m **c.** +1.0 m **d.** +3.0 m
8. The teacher's average velocity for the time interval 1.0–3.0 s is _____.
   a. –1.0 m/s **b.** +1.0 m/s **c.** +2.0 m/s **d.** +3.0 m/s
9. The teacher is standing still during the time interval _____.
   a. 1.0–2.0 s **b.** 6.0–7.0 s **c.** 10.0–12.0 s **d.** 12.0–15.0 s
10. The average velocity for the time interval 10.0–12.0 s is _____.
   a. +1.0 m/s **b.** +0.5 m/s **c.** 0.0 m/s **d.** –0.5 m/s
11. The teacher's average velocity for the time interval 0.0–15.0 s is _____.
   a. –2.0 m/s **b.** 0.0 m/s **c.** +0.067 m/s **d.** +0.75 m/s

Sketch a motion diagram for each time interval, using the teacher's position-time graph on page 26.

	Time Interval	*Motion Diagram*
12.	0.0–5.0 s	
13.	5.0–7.0 s	
14.	7.0–15.0 s	

In your textbook, read about using position and time equations.
Refer to the equation $d = d_0 + vt$. For each of the statements below, write true *or* rewrite *the italicized part to make the statement true.*

15. ______________________ The equation contains *three* quantities.

16. ______________________ This equation *cannot* be used if the velocity is changing.

17. ______________________ The quantity d_0 represents the position at any time.

18. ______________________ The quantity in the equation that represents the slope of a position-time graph for this motion is v.

19. ______________________ The displacement of the object is vt.

Name ______________________________

5 Study Guide

Section 5.2: Graphing Velocity in One Dimension

In your textbook, read about velocity-time graphs and displacement.

Refer to the velocity-time graph of a jogger to complete the two tables.

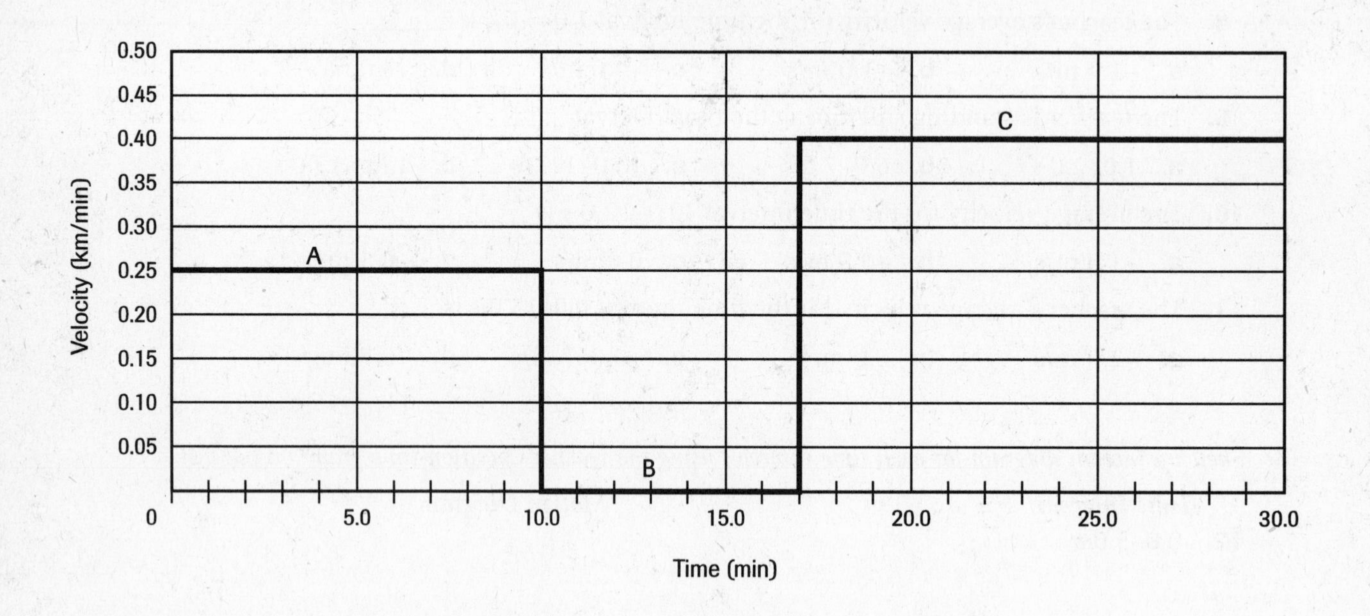

Table 1

Segment	v	Δt	Δd
A			
B			
C			

Table 2

Δt	Distance Ran	Displacement	Average Velocity

Name ______________________

5 Study Guide

Section 5.3: Acceleration

In your textbook, read about equations of motion for uniform acceleration.

Complete each table with the values of the variables and initial conditions. Write a question mark for the unknown variable. If a variable or an initial condition is not needed to answer the problem, write an X. Then write the equation that you would use to solve the problem. It is not necessary to calculate the answer.

1. A car accelerates from 10 m/s to 15 m/s in 3.0 s. How far does the car travel in this time?

Table 1						
Variables			**Initial Conditions**			**Equation**
t	d	v	a	d_0	v_0	

2. A racing car accelerates at 4.5 m/s^2 from rest. What is the car's velocity after it has traveled 35.0 m?

Table 2						
Variables			**Initial Conditions**			**Equation**
t	d	v	a	d_0	v_0	

3. A car initially traveling at 15 m/s accelerates at a constant 4.5 m/s^2 through a distance of 45 m. How long does it take the car to cover this distance?

Table 3						
Variables			**Initial Conditions**			**Equation**
t	d	v	a	d_0	v_0	

4. A ball rolls past a mark on an incline at 0.40 m/s. If the ball has a constant acceleration of 0.20 m/s^{2}what is its velocity 3.0 s after it passes the mark?

Table 4						
Variables			**Initial Conditions**			**Equation**
t	d	v	a	d_0	v_0	

5 Study Guide

Name ______________________

Section 5.4: Free-Fall

In your textbook, read about free-fall.

Refer to the diagram on the right showing the positions of a ball that was thrown upward at time t_0. Complete the tables by indicating if the direction of the velocity and the direction of the acceleration at each time is + or −, or if the value of the velocity or acceleration equals 0.

O t_2

O t_3

O t_1

O t_4

O t_0

1. Downward direction is positive.

Rank the magnitudes of the velocities v_0, v_1, v_2, v_3, and v_4, in decreasing order.

____

____

Table 1

Variable	Time t_0	t_1	t_2	t_3	t_4
v					
a					

2. Upward direction is positive.

Rank the magnitudes of the velocities v_0, v_1, v_2, v_3, and v_4, in decreasing order.

____

____

Table 2

Variable	Time t_0	t_1	t_2	t_3	t_4
v					
a					

Refer to the velocity-time graph of the vertical velocity of a ball from the time it is thrown upward from the top of a building until it reaches the ground. Answer the following questions.

3. Use a red pencil to shade the area under the curve that represents the total upward displacement of the ball.

4. Use a blue pencil to shade the area under the curve that represents the total downward displacement of the ball.

5. How could you use the red and blue areas to determine the height of the building?

____

____

____

____

____

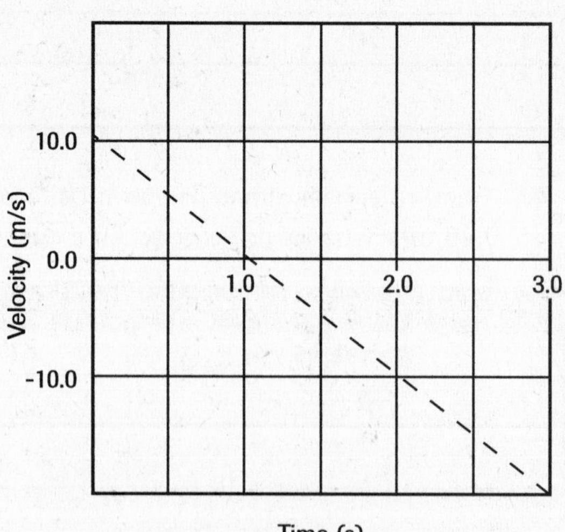

Date ______________ Period ______________ Name ______________________________

6 Study Guide

Use with Chapter 6.

Forces

Vocabulary Review

Write the term that correctly completes each statement. Use each term once.

agent	force of gravity	mechanical resonance	simple harmonic motion
amplitude	free-body diagram	net force	static friction force
apparent weight	inertia	Newton's first law	system
contact force	interaction pair	Newton's second law	terminal velocity
environment	kinetic friction force	Newton's third law	weightlessness
equilibrium	long-range force	period	
force			

1. ______________________ In _____, the amplitude of motion is increased by the repeated application of a force when the time between applications is equal to the period of oscillation.
2. ______________________ The vector sum of two or more forces acting on an object is the _____.
3. ______________________ The net force on an object in _____ is zero.
4. ______________________ The horizontal force exerted on one surface by another when the two surfaces are not in relative motion is the _____.
5. ______________________ The acceleration of a body is directly proportional to the net force on it and inversely proportional to its mass; this is a statement of _____.
6. ______________________ The force exerted by a scale on an object is the _____.
7. ______________________ A force that acts on an object by touching it is a(n) _____.
8. ______________________ The two forces in an interactive pair act on different objects and are equal in magnitude and opposite in direction; this is a statement of _____.
9. ______________________ The world outside a system is the _____.
10. ______________________ An attractive force that exists between all objects is the _____.
11. ______________________ An object that is at rest will remain at rest or an object that is moving will continue to move in a straight line with constant speed, if and only if the net force acting on the object is zero; this is a statement of _____.
12. ______________________ The horizontal force exerted on one surface by another when surfaces are in relative motion is the _____.
13. ______________________ A defined object or group of objects is a(n) _____.

Name ______________________________

6 Study Guide

14. ______________________ The tendency of an object to resist changes in its motion is _____.
15. ______________________ The specific, identifiable cause of a force is a(n) _____.
16. ______________________ The time to repeat one complete cycle is a(n) _____.
17. ______________________ A force that is exerted without contact is a(n) _____.
18. ______________________ In any periodic motion, the maximum displacement from equilibrium is the _____.
19. ______________________ A push or pull is a(n) _____.
20. ______________________ Two forces that are in opposite directions and have equal magnitudes are a(n) _____.
21. ______________________ In a(n) _____, a dot represents an object and arrows represent each force acting on it, with their tail on the dot and their points indicating the direction of the force.
22. ______________________ In _____, the force that restores an object to its equilibrium position is directly proportional to its displacement.
23. ______________________ The constant velocity that a falling object reaches when the drag force equals the force of gravity is its _____.
24. ______________________ When an object's apparent weight is zero, the object is in a state of _____.

Section 6.1: Force and Motion

In your textbook, read about Newton's second law and combining forces.

For each of the statements below write true *or* false.

__________ **1.** Newton's second law can be written as the equation

$$a = \frac{F_{net}}{m}.$$

__________ **2.** The acceleration of an object and the net force acting on it are proportional.

__________ **3.** Force and acceleration are scalar quantities.

Name ______________________

6 Study Guide

In your textbook, read about equilibrium and free-body diagrams.

Refer to the diagrams below. Circle the letter of the choice that best completes the statement or answers the question.

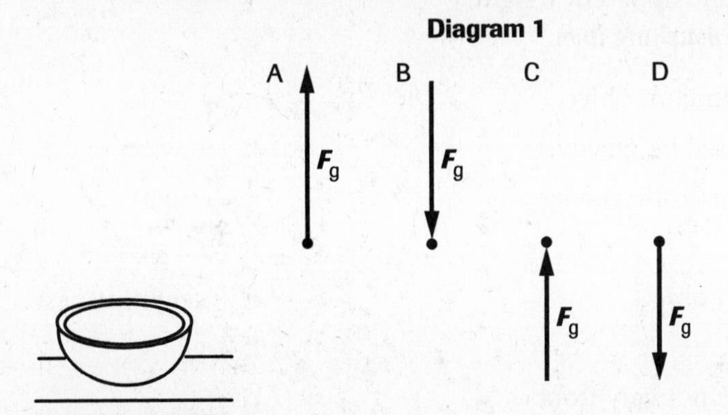

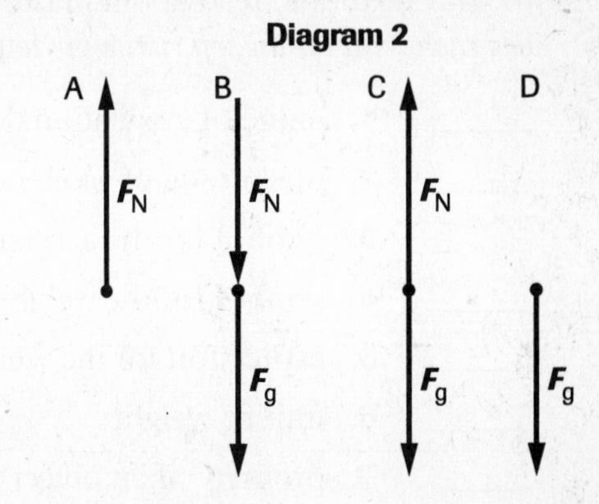

4. Which part of diagram 1 best represents the weight force of the bowl sitting on a shelf?

a. A **b.** B **c.** C **d.** D

5. F_N is a symbol that represents the _____ force.

a. friction **b.** tension **c.** normal **d.** weight

6. What part of diagram 2 best represents the bowl in equilibrium?

a. A **b.** B **c.** C **d.** D

7. The magnitude of the net force on the bowl in equilibrium is _____.

a. F_N **b.** F_g **c.** 0 **d.** $2F_g$

8. The agent of F_N is _____.

a. the bowl **b.** Earth **c.** friction **d.** the shelf

9. The agent of F_g is _____.

a. the bowl **b.** Earth **c.** friction **d.** the shelf

Draw a free-body diagram of each situation.

10. A rocket immediately after vertical liftoff

11. A penny sliding at constant velocity on a desktop

12. A penny immediately after sliding off a desktop

Name ______________________________

6 Study Guide

Section 6.2: Using Newton's Laws

In your textbook, read about mass, weight, and apparent weight.
For each term on the left, write the letter of the matching item.

_____ **1.** name of gravitational force acting on object

_____ **2.** magnitude of acceleration caused by gravity

_____ **3.** symbol for the acceleration caused by gravity

_____ **4.** symbol for the weight force

_____ **5.** expression for the weight of an object

_____ **6.** unit of weight

_____ **7.** property of an object that does not vary from location to location

_____ **8.** having an apparent weight of zero

a. g
b. newton
c. weight
d. 9.8 m/s^2
e. weightlessness
f. mg
g. F_g
h. mass

In your textbook, read about scales and apparent weight.
Read the passage below and refer to the diagram next to it. Circle the letter of the choice that best completes the statement or answers the question.

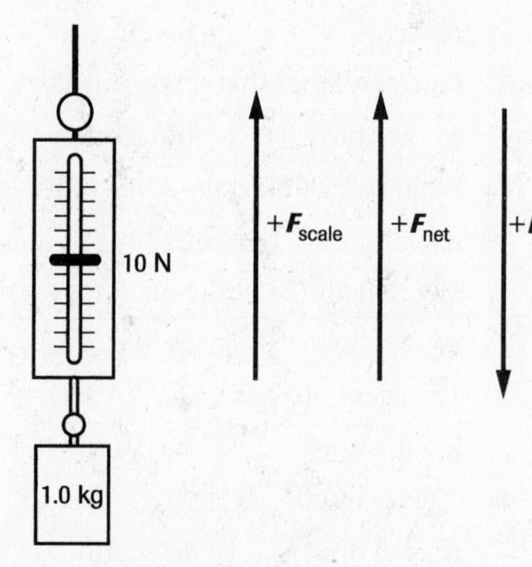

A 1.0-kg mass at rest is suspended from a spring scale. The direction of positive forces that are acting or could act on the 1.0-kg mass are shown to the right.

9. The 1.0-kg mass and spring scale are being lifted at a constant speed. The net force on the mass is _____.

a. 0 N **c.** −10 N
b. +10 N **d.** +20 N

10. The 1.0-kg mass and spring scale are being lifted so that the 1.0-kg mass is being accelerated in the positive upward direction at 1.0 m/s^2. What is the net force acting on the mass?

a. 0 N **b.** +1 N **c.** −1 N **d.** +20 N

11. In problem 10, what is the relationship among the magnitudes of the forces acting on the mass?

a. $F_{net} = F_{scale} + F_g$ **c.** $F_{net} = -(F_{scale} + F_g)$
b. $F_{net} = F_{scale} - F_g$ **d.** $F_{net} = F_g - F_{scale}$

12. In problem 10, what is the spring scale reading?

a. < 10 N **b.** 10 N **c.** >10 N **d.** 0 N

13. If the scale is accidentally dropped, the net force acting on the 1.0-kg mass is _____.

a. 0 N **b.** +10 N **c.** −10 N **d.** +20 N

14. If the scale is accidentally dropped, the reading of the spring scale as it falls is _____.

a. 0 N **b.** +10 N **c.** −10 N **d.** +20 N

Name ______________________________

6 Study Guide

In your textbook, read about the friction force.

For each of the statements below, write true *or rewrite the italicized part to make the statement true.*

15. ______________________ The force of friction *opposes* the motion of an object sliding on a surface.

16. ______________________ The force of friction acts in a direction that is *perpendicular* to the surface on which the object slides.

17. ______________________ The force of friction acts in the *same* direction as the direction of the sliding object.

18. ______________________ Static friction opposes the *start* of an object's motion.

19. ______________________ The static friction force *does* depend on the area of the surfaces in contact.

20. ______________________ The kinetic friction force *does* depend on the speed of the relative motion of the surfaces.

In your textbook, read about periodic motion.

Read the sentence below and answer the following questions, using complete sentences.

A stone attached to a spring is oscillating vertically.

21. What is the direction of the stone's acceleration when it is at the lowest point of its oscillation?

____

____

22. At what point in its oscillation does the stone have the greatest downward acceleration?

____

____

23. At what point does the stone have an acceleration of zero in its downward motion?

____

____

24. At what point does the stone have its maximum upward and downward velocities?

____

____

25. What causes the stone to move through the midpoint of its oscillation?

____

____

Name ______________________________

6 Study Guide

Section 6.3: Interaction Forces

In your textbook, read about interaction pairs.
Refer to the diagram below to complete the table.

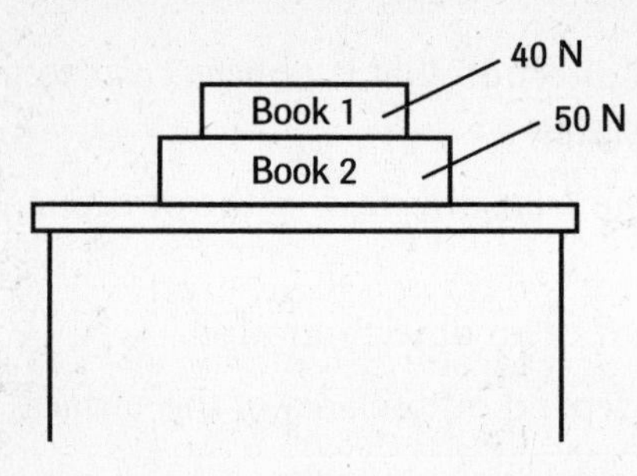

Table 1		
Force	**Magnitude**	**Direction**
$F_{\text{book 1 on book 2}}$		
$F_{\text{book 2 on book 1}}$		
$F_{\text{book 2 on desktop}}$		
$F_{\text{desktop on book 2}}$		
$F_{\text{books 1 and 2 on desktop}}$		
$F_{\text{desktop on books 1 and 2}}$		

In your textbook, read about the four fundamental forces.
For each answer given below, write an appropriate question.

1. Answer: an attractive force due to the masses of objects
   Question: ______________________________

2. Answer: the electromagnetic interaction
   Question: ______________________________

3. Answer: two forces that occur within the nucleus of the atom
   Question: ______________________________
   ____

4. Answer: the strong nuclear interaction
   Question: ______________________________
   ____

5. Answer: the force apparent in some kinds of radioactive decay
   Question: ______________________________

Date ______________ Period ______________ Name ______________________

7 Study Guide

Use with Chapter 7.

Forces and Motion in Two Dimensions

Vocabulary Review

Write the term that correctly completes each statement. Use each term once.

centripetal acceleration	**lever arm**	**range**
centripetal force	**maximum height**	**rigid rotating object**
coefficient of kinetic friction	**normal force**	**torque**
equilibrant	**period**	**trajectory**
equilibrium	**projectile**	**uniform circular motion**
flight time		

1. ______________________ The perpendicular distance from the axis of rotation to a line along which a force acts is the _____.
2. ______________________ The path of a projectile is the _____.
3. ______________________ A force that produces equilibrium is the _____.
4. ______________________ The horizontal distance traveled by a projectile is the _____.
5. ______________________ A force that acts toward the center of a circle is the _____.
6. ______________________ The height of a projectile when its vertical velocity is zero is the _____.
7. ______________________ A mass that rotates around its own axis is a(n) _____.
8. ______________________ The time a projectile is in the air is the _____.
9. ______________________ Acceleration toward the center of a circle is the _____.
10. ______________________ An object that has been launched by an initial thrust is a(n) _____.
11. ______________________ A contact force that acts perpendicular to and away from a surface is a(n) _____.
12. ______________________ The product of a force and the lever arm is the _____.
13. ______________________ The movement of an object or a point mass at constant speed around a circle with a fixed radius is _____.
14. ______________________ When the net sum of the forces acting on an object is zero, the object is in _____.
15. ______________________ The ratio of the force of kinetic friction to the normal force is the _____.
16. ______________________ The time needed to complete one cycle of motion is a(n) _____.

Name ______________________

7 Study Guide

Section 7.1: Forces in Two Dimensions

In your textbook, read about the equilibrant.

For each of the statements below, write true *or rewrite the italicized part to make the statement true.*

1. ______________________ Equilibrium occurs when the resultant of two or more forces equals a net force *greater than* zero.

2. ______________________ An object in equilibrium is motionless or has a *constant velocity*.

3. ______________________ The *equilibrant* is a force that places an object in equilibrium when added to the resultant of all the forces on the object.

4. ______________________ The equilibrant is *equal* in magnitude to the resultant force.

5. ______________________ The equilibrant acts in the *same direction* as the resultant force.

Refer to the diagram below. Circle the letter of the choice that best completes each statement.

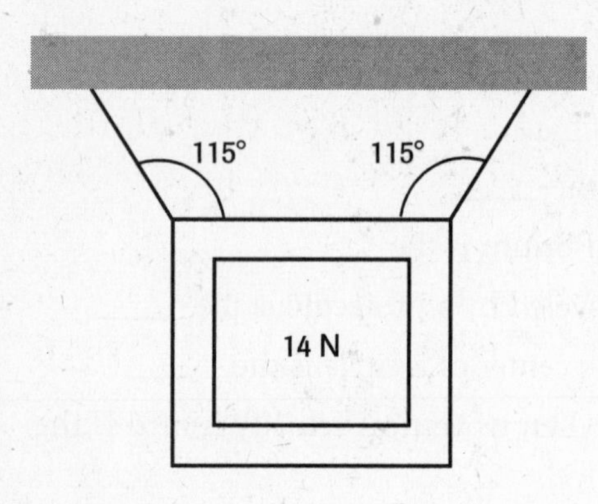

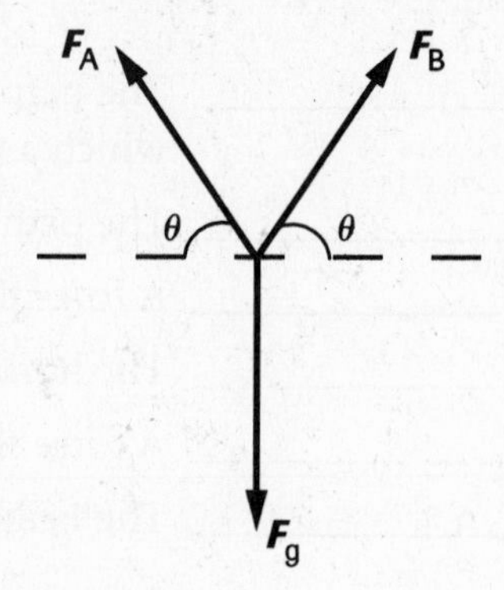

6. Angle θ is _____.

a. 57.5° **b.** 65° **c.** 115° **d.** 230°

7. The force equation that represents equilibrium is _____.

a. $F_A - F_B - F_g = 0$

b. $F_A = 2F_g$

c. $F_A = F_B$

d. $F_A + F_B - F_g = 0$

8. $F_A \cos \theta =$ _____.

a. $F_A \sin \theta$ **b.** $-F_B \cos \theta$ **c.** F_g **d.** $F_g/2$

9. $F_B \sin \theta =$ _____.

a. $F_A \cos \theta$ **b.** $F_B \cos \theta$ **c.** F_g **d.** $F_g/2$

10. As θ increases, the tension on each rope _____.

a. increases **b.** decreases **c.** remains the same

Name ______________________

7 Study Guide

In your textbook, read about motion along an inclined plane.

Draw a free-body diagram representing the situation below. Label angle θ and the horizontal and vertical components of the weight force.

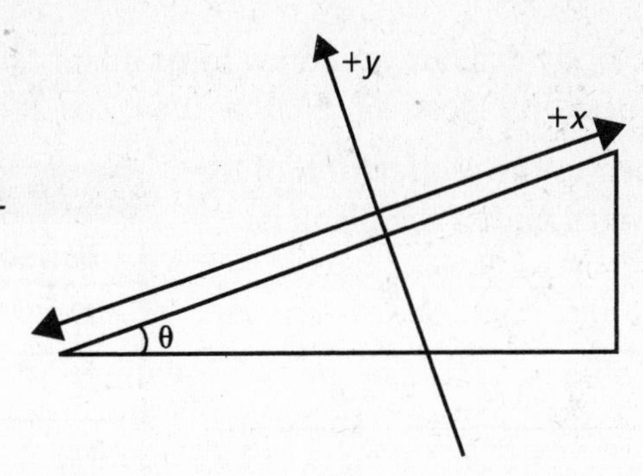

A trunk is placed on a frictionless inclined plane that is at angle θ from the horizontal.

For each situation below, write the letter of the matching item.

______ **11.** calculation of F_{gx}

______ **12.** calculation of F_{gy}

______ **13.** calculation of F_N

______ **14.** calculation of F_g

______ **15.** component of weight force that increases as angle θ increases

______ **16.** component of weight force that decreases as angle θ increases

______ **17.** calculation of acceleration along incline

a. perpendicular component of weight force

b. mg

c. $-F_g \cos \theta$

d. $-g \sin \theta$

e. $-F_g \sin \theta$

f. parallel component of weight force

g. $-F_{gy}$

18. Draw a free-body diagram representing the situation below. Label the $+x$-axis in the direction of motion. Label the $+y$-axis. Then draw and label items a–f.

An 82-N trunk is accelerating along a plank inclined at 26.0° from the horizontal. The coefficient of kinetic friction of the trunk and plank is 0.20.

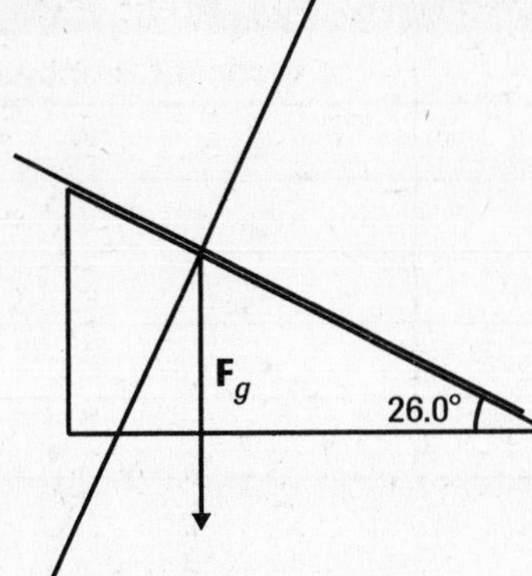

a. horizontal component of the weight force

b. vertical component of the weight force

c. angle of 26.0°

d. normal force

e. force of kinetic friction

f. net force

Name ______________________

7 Study Guide

19. Refer to the free-body diagram you drew in problem 18. Complete the table below. Then answer the following questions. Note: sin 26.0° = 0.44; cos 26.0° = 0.90.

Table 1			
Variable	**Value**	**Variable**	**Value**
F_g		F_N	
F_{gx}		F_f	
F_{gy}		F_{net}	

20. What will happen to the acceleration of the trunk if the elevated end of the plank is raised? Why?

____

____

____

____

____

21. What will happen to the motion of the trunk if it reaches a section of the plank where the coefficient of sliding friction makes the frictional force equal to the magnitude of the horizontal component of the weight force?

____

Section 7.2: Projectile Motion

In your textbook, read about solving projectile-motion problems.

Read each problem and complete the table by writing the value of the variable, an X if the variable is not necessary to answer the problem, or a question mark for the unknown. (Assume that the coordinate system has the +y-axis pointing up.) Briefly explain how you would solve the problem.

1. A ball rolls off a horizontal shelf at 2.0 m/s and falls to the floor 1.8 m below. How far from the shelf does the ball land?

Table 2		
Variable	**Vertical Component**	**Horizontal Component**
acceleration $a = F_{net}/m$	$a_y =$	$a_x =$
velocity $v = v_0 + at$	$v_{y0} =$	$v_{x0} =$
	$v_y =$	$v_x =$
position $d = d_0 + v_0t + \frac{1}{2}at^2$	$y_0 =$	$x_0 =$
	$y =$	$x =$

____

____

____

____

7 Study Guide

2. A javelin leaves the thrower's hand at a velocity of 21 m/s at an angle of 43° above the horizontal. What is the maximum height the javelin will reach? Note: (21 m/s)(sin 43°) = 14 m/s; (21 m/s)(cos 43°) = 15 m/s.

Table 3

Variable	Vertical Component	Horizontal Component
acceleration $a = F_{net}/m$	$a_y =$	$a_x =$
velocity $v = v_0 + at$	$v_{y0} =$	$v_{x0} =$
	$v_y =$	$v_x =$
position $d = d_0 + v_0t + \frac{1}{2}at^2$	$y_0 =$	$x_0 =$
	$y =$	$x =$

____

____

____

____

3. A person at the edge of a vertical 12-m cliff throws a rock down at a velocity of 8.5 m/s, 55° below the horizontal. How far from the base of the cliff does the rock land? Assume a coordinate system in which the upward direction is positive. Note: (8.5 m/s)(sin 55°) = 7.0 m/s; (8.5 m/s)(cos 55°) = 4.9 m/s.

Table 4

Variable	Vertical Component	Horizontal Component
acceleration $a = F_{net}/m$	$a_y =$	$a_x =$
velocity $v = v_0 + at$	$v_{y0} =$	$v_{x0} =$
	$v_y =$	$v_x =$
position $d = d_0 + v_0t + at^2$	$y_0 =$	$x_0 =$
	$y =$	$x =$

____

____

____

____

Name ______________________________

7 Study Guide

Section 7.3: Circular Motion

In your textbook, read about circular motion.

Write the term that correctly completes each statement. Use each term once.

center	**directions**	**length**	**straight line**
centripetal	**directly**	**perpendicular**	**tangent**
constant	**inversely**	**radius**	

An object is moving in uniform circular motion if the radius of its path and its speed are **(1)** ______________. The velocity of an object in uniform circular motion is **(2)** ______________ to the radius and **(3)** ______________ to the circular path. The velocity vectors have the same **(4)** ______________ but different **(5)** ______________. The direction of the centripetal acceleration vector lies along the **(6)** ______________ and points toward the **(7)** ______________. The centripetal acceleration is **(8)** ______________ proportional to the square of the velocity and **(9)** ______________ proportional to the radius of the circle. The force that causes the centripetal acceleration is the **(10)** ______________ force. If this force ceases to act on the object in uniform circular motion, the object will travel in a **(11)** ______________ tangent to the circle.

In your textbook, read about torque.

The diagrams below represent a force of 65 N applied at three different angles to a wrench 0.25 m long and connected to the head of a bolt. In each diagram, construct the lever arm and label it. Then answer the following questions.

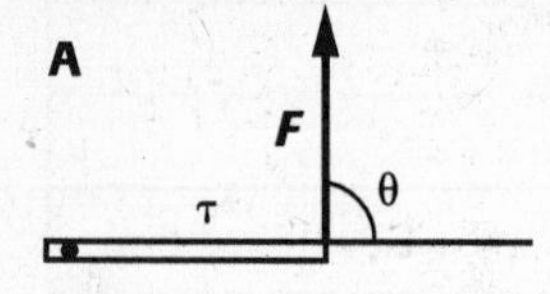

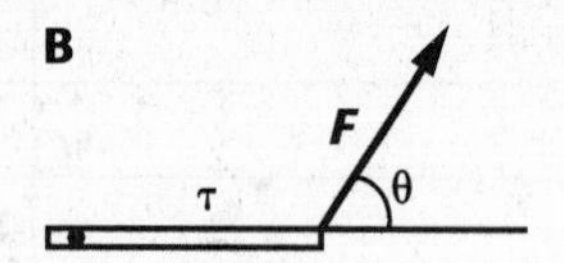

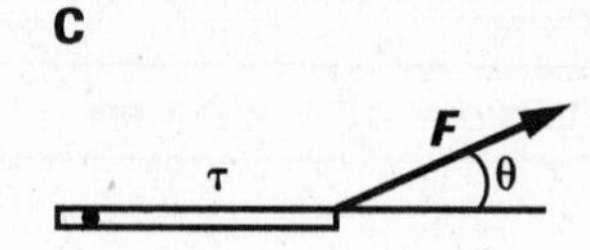

12. In which diagram is the torque the greatest?

____

13. At what angle θ will the torque be maximal?

____

14. At what angle θ will the torque be minimal?

____

Date ______________ Period ______________ Name ______________________________

8 Study Guide

Use with Chapter 8.

Universal Gravitation

Vocabulary Review

Write the term that correctly completes each statement. Use each term once.

black hole	**inertial mass**	**law of universal gravitation**
centripetal acceleration	**inverse square law**	**Newton's universal gravitational constant**
gravitational field	**Kepler's laws of planetary motion**	**period**
gravitational mass		

1. ______________________ The behavior of every planet and satellite is described by _____.

2. ______________________ The _____ states that the gravitational force between two objects depends directly on the product of their masses and inversely on the square of the distance between the centers of the masses.

3. ______________________ The _____ of an object is $\frac{F_{net}}{a}$, where F_{net} is the net force exerted on an object and a is its acceleration.

4. ______________________ The _____ of an object is defined as $\frac{r^2 F_{grav}}{Gm}$ where F_{grav} is the gravitational force exerted on an object by another object of mass m at a distance r, and G is a universal constant.

5. ______________________ The _____ is a property of the region surrounding an object, is due to the object's mass, and interacts with other objects, resulting in a force of attraction.

6. ______________________ A massive and dense object in which light leaving the object is totally bent back to the object, resulting in no light escaping is a(n) _____.

7. ______________________ The time needed to repeat one complete cycle of motion is a(n) _____.

8. ______________________ The direction of _____ is toward the center of motion.

9. ______________________ A relationship in which one variable is inversely proportional to the square of another variable is the basis of a(n) _____.

10. ______________________ _____ is the same number for any two masses anywhere.

Name ____________________

8 Study Guide

Section 8.1: Motions in the Heavens and on Earth

In your textbook, read about observed motion and Kepler's laws.

For each statement below, write true *or rewrite the italicized part to make the statement true.*

1. Tycho Brahe believed that Earth was the center of the universe.

____

2. Johannes Kepler believed *the sun* was the center of the universe.

____

3. From studying Brahe's astronomical data, Kepler developed three *physical* laws of planetary motion.

____

4. Kepler's first law states that the paths of the planets are *ellipses.*

____

5. In the first law, *the planet* is located at one focus of the planet's orbit.

____

6. The second law states that an imaginary line extending from the sun to the planet sweeps out equal *distances* in equal times.

____

7. The third law states that the *product* of the squares of the periods of any two planets revolving about the sun is equal to the cube of the ratio of their average distances from the sun.

____

8. Kepler's *first, second, and third* laws apply to each planet, moon, or satellite individually.

____

In your textbook, read about Kepler's laws.

Refer to the diagram to answer the following questions.

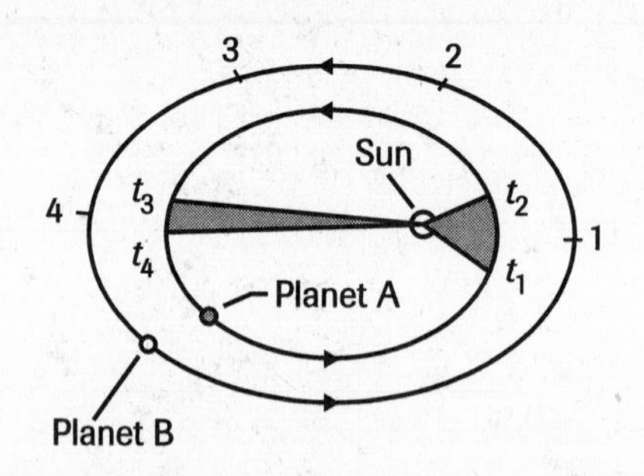

9. The shaded portions of planet A's orbit represent areas swept out by an imaginary line from the sun between times t_1 and t_2 and between times t_3 and t_4. If the time intervals $t_2 - t_1$ and $t_4 - t_3$ are equal, what statement can be made about the two shaded areas?

____

____

Name ______________________

8 Study Guide

10. At which of the indicated points in planet B's orbit is its speed the greatest?

____

____

11. Io, one of Jupiter's moons, has a period of 1.8 days. What information would you need to apply Kepler's third law to determine how far Io is from the center of Jupiter?

____

____

____

In your textbook, read about observed motion and universal gravitation.
For each term on the left, write the letter of the matching item.

_____ **12.** German mathematician who developed three laws of planetary motion

_____ **13.** Danish astronomer who recorded the exact positions of the planets and stars for over 20 years

_____ **14.** English physicist who proposed that gravitational attraction keeps planets and satellites in their orbits

_____ **15.** Italian scientist who discovered the moons of Jupiter

a. Brahe
b. Galileo
c. Kepler
d. Newton

In your textbook, read about universal gravitation.
Refer to the diagram and complete the table for the new values of distance and mass.

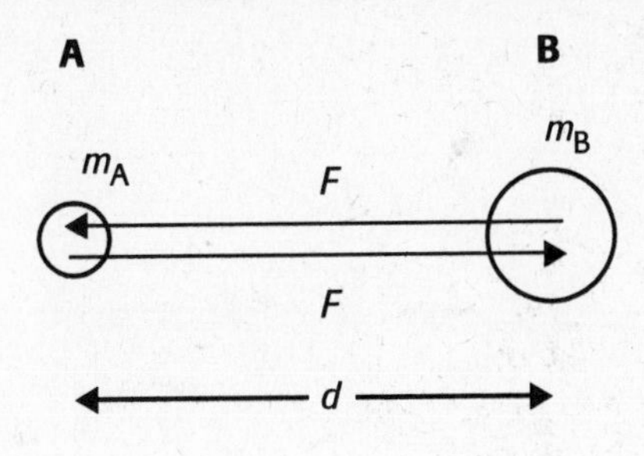

Table 1

Distance	Mass A	Mass B	Force
16. d	m_A	$2m_B$	
17. d	$2m_A$	m_B	
18. d	$2m_A$	$2m_B$	
19. $2d$	m_A	m_B	
20. $d/2$	m_A	m_B	

Name ______________________

8 Study Guide

Read about universal gravitation and weighing the earth.

Answer the following questions, using complete sentences.

21. What apparatus did Cavendish use to find a value for Newton's universal gravitational constant?

____

____

____

____

22. How did Cavendish get an experimental value for gravitational force, F, to use in Newton's law of gravitational force?

____

____

____

____

____

23. The gravitational force between two 1.00-kg masses whose centers are 1.00 m apart is defined as 6.67×10^{-11} N. Show that the value of Newton's universal gravitational constant is 6.67×10^{-11} $N \cdot m^2/kg^2$.

Name ______________________

8 Study Guide

Section 8.2: Using the Law of Universal Gravitation

In your textbook, read about the motion of planets and satellites.
Write the term that correctly completes each statement.

If a cannonball is fired horizontally from a cannon high atop a mountain, its **(1)** ______________________ has both horizontal and vertical components. During the first second of its flight, the cannonball falls **(2)** ______________________. If fired at the right horizontal velocity, the cannonball will fall 4.9 m at a point where Earth's surface has curved 4.9 m from the **(3)** ______________________. Thus, the cannonball is at the same **(4)** ______________________ above Earth as it was initially. The curvature of the cannonball's path will continue to just match the **(5)** ______________________ of Earth. As a result, the cannonball is said to be in **(6)** ______________________. A horizontal speed of 8 km/s will keep the cannonball at the same altitude, and it will circle Earth as an artificial **(7)** ______________________.

In your textbook, read about the gravitational field.
Refer to the diagram on the right to answer the following questions.

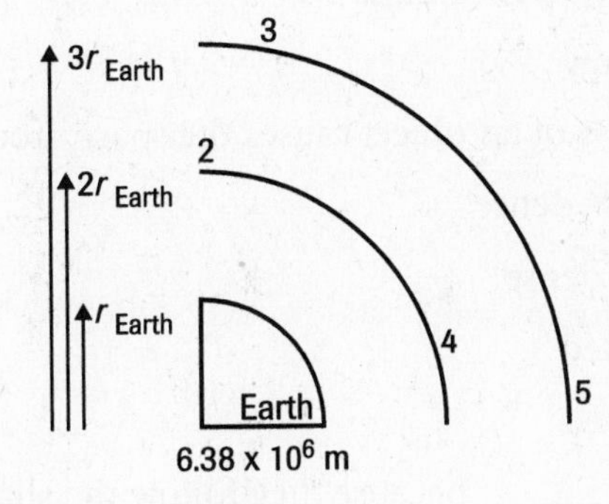

8. Draw the vector representing the direction and magnitude of the gravitational field at each of the points 1–5. Indicate the scale you used. Label each gravitational field vector with its magnitude.

9. What is the weight of a 10.0-kg mass located at point 4?

____

In your textbook, read about two kinds of masses.
For each term on the left, write the letter of the matching item.

____	**10.** definition of inertial mass	**a.**	resistance to acceleration
____	**11.** definition of gravitational mass	**b.**	principle of equivalence
____	**12.** property of inertial mass	**c.**	$\frac{F_{net}}{a}$
____	**13.** property of gravitational mass	**d.**	attraction toward other masses
____	**14.** hypothesis that inertial and gravitational masses are identical	**e.**	$\frac{r^2 F_{grav}}{Gm}$

Name ____________________

8 Study Guide

In your textbook, read about Einstein's theory of gravity.

Circle the letter of the choice that best completes each statement.

15. The effect of gravity can be explained by the _____

a. equivalence principle

b. gravitational field

c. general theory of relativity

d. inertial field

16. The origin of gravity is proposed in _____.

a. the law of universal gravitation

b. the gravitational field

c. the general theory of relativity

d. the inertial field

17. According to Einstein, gravity is _____.

a. an effect of space

b. a force

c. a property of mass

d. energy

18. The mass of an object causes the space around it to become _____.

a. more dense

b. less dense

c. heated

d. curved

19. Objects _____ because they follow the shape of the space surrounding a massive object.

a. are repulsed

b. become visible

c. lose mass

d. undergo accelerations

20. The effect of mass on space _____.

a. accounts for the deflection of light by massive objects

b. explains the existence of black holes

c. models gravitational attraction

d. all of these

Date ______________ Period ______________ Name ______________________

9 Study Guide

Use with Chapter 9.

Momentum and Its Conservation

Vocabulary Review

Write the term that correctly completes each statement. Use each term once.

angular momentum	impulse-momentum theorem	law of conservation of momentum	Newton's third law
closed system	internal forces	linear momentum	recoil
external forces	isolated system	Newton's second law	torque
impulse			vector

1. ______________ The acceleration of a body is directly proportional to the net force on it and inversely proportional to its mass; this is a statement of _____.
2. ______________ When one object exerts a force on a second object, the second exerts a force on the first that is equal in magnitude but opposite in direction; this is a statement of _____.
3. ______________ A quantity that has magnitude and direction is a(n) _____.
4. ______________ The product of the average force exerted on an object and the time interval through which it acts is the _____.
5. ______________ The product of the mass and velocity of a body is its _____.
6. ______________ The _____ states that the impulse on an object is equal to the change of momentum that it causes.
7. ______________ The quantity of motion used with objects rotating about a fixed axis is _____.
8. ______________ A(n) _____ does not gain or lose mass.
9. ______________ Forces within a closed system are _____.
10. ______________ All the forces that act on the system from outside the system are _____.
11. ______________ A closed system in which the net external force on the system is zero is a(n) _____.
12. ______________ The _____ states that the momentum of any closed system with no net external force acting on it does not change.
13. ______________ The backward motion of part of an isolated, closed system due to an internal force is _____.
14. ______________ The product of force and a lever arm is _____.

Name ______________________

9 Study Guide

Section 9.1: Impulse and Momentum

In your textbook, read about relating impulse and momentum.

The diagram below shows the force-time graph of a force acting on an 8.0-kg cart initially at rest on a frictionless surface. Circle the letter of the choice that best completes each statement.

Force (N): 20.0, 18.0, 16.0, 14.0, 12.0, 10.0, 8.0, 6.0, 4.0, 2.0, 0.0

Time (s): 0.0 1.0 2.0 3.0 4.0 5.0 6.0 7.0 8.0 9.0 10.

1. The magnitude of the force acting on the cart is _____.

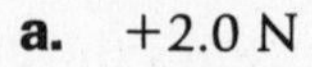

a. +2.0 N **c.** +10.0 N

b. +8.0 N **d.** +20.0 N

2. The force acted on the cart for a duration of _____.

a. 2.0 s **c.** 8.0 s

b. 6.0 s **d.** 10.0 s

3. The shaded area of the graph represents _____.

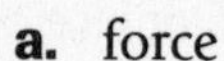

a. force **c.** momentum

b. impulse **d.** acceleration

4. The shaded area of the graph is represented algebraically by _____.

a. $F\Delta t$ **c.** $F/\Delta t$

b. mv **d.** m/v

5. The impulse acting on the cart is _____.

a. 0 N·s **c.** +8.0 N·s

b. +2.0 N·s **d.** +48 N·s

6. The momentum of the cart is represented algebraically by _____.

a. $F\Delta t$ **c.** $F/\Delta t$

b. mv **d.** m/v

7. The change of momentum of the cart is _____.

a. 0 kg·m/s **c.** +8.0 kg·m/s

b. +2.0 m/s **d.** +48 kg·m/s

8. The final velocity of the cart is _____.

a. 0 m/s **c.** 12 m/s

b. 6.0 m/s **d.** 48 m/s

Name ____________________

9 Study Guide

In your textbook, read about using the impulse-momentum theorem.

Read each of the statements below. In the diagrams, draw and label the velocities of the sport utility van before and after the event described. In the vector diagram section, draw and label the momentum vectors $\mathbf{p}_1$ and $\mathbf{p}_2$, and the impulse vector $\mathbf{F}\Delta t$.

9. The van has a velocity of +30 km/h and uniformly comes to a stop in 15 s.

Before ***After*** ***Vector diagram***

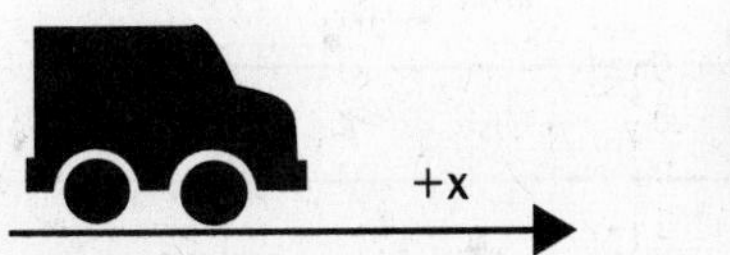

10. The van, initially at rest, uniformly reaches a velocity of +30 km/h in 20 s.

Vector diagram

11. In 5 s, the van uniformly backs up to a speed of 5 km/h.

Vector diagram

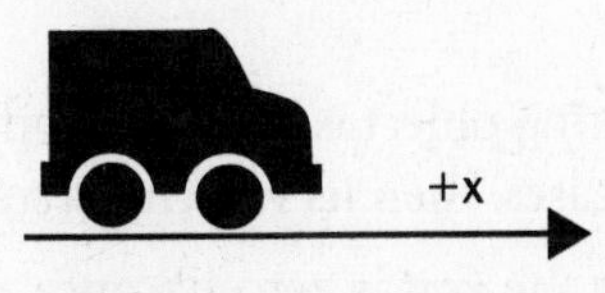

12. The van has a velocity of −5 km/h and uniformly comes to a stop in 5 s.

Vector diagram

Name ______________________________

9 Study Guide

In your textbook, read about the impulse-momentum theorem as it applies to air bags.
Answer the following questions, using complete sentences.

13. In what two ways can impulse be increased?

____

____

14. In terms of impulse, how does an air bag reduce injuries that would be caused by the steering wheel during an accident?

____

____

____

____

____

____

In your textbook, read about angular momentum.
For each of the statements below, write true *or rewrite the italicized part to make the statement true.*

15. ______________________ The speed of a rotating object changes only if *torque* is applied to it.

16. ______________________ The quantity of angular motion that is used with rotating objects is called *linear* momentum.

17. ______________________ When a *torque* acts on an object, its angular momentum changes.

18. ______________________ Angular momentum is the product of the object's mass, displacement from the center of rotation, and the component of its velocity *parallel* to that displacement.

19. ______________________ If the angular momentum of a rotating object is constant and its distance to the center of rotation decreases, then its velocity *decreases.*

20. ______________________ The torque on the planets orbiting the sun is zero because the gravitational force is directed toward the sun and, therefore, is not *perpendicular* to its velocity.

21. ______________________ Because the torque on each planet is zero, each planet's angular momentum is *zero.*

22. ______________________ When a planet's distance from the sun becomes smaller, the planet moves *slower.*

Name ____________________

9 Study Guide

Section 9.2: The Conservation of Momentum

In your textbook, read about two-particle collisions.

Read the sentence below and answer the following questions, using complete sentences.

Two balls of unequal mass traveling at different speeds collide head-on and rebound in opposite directions.

1. How does the force that ball A exerts on ball B compare to the force that ball B exerts on ball A?

____

2. How do the impulses received by both balls compare?

____

3. How do the sums of the momenta of the balls before and after the collision compare?

____

Write the term that correctly completes each statement. Use each term once.

change	**conditions**	**interaction**	**isolated system**
closed system	**external forces**	**internal forces**	**law of conservation of momentum**

A system that doesn't gain or lose mass is a(n) **(4)** ____________________. All the forces within a closed system are **(5)** ____________________. All the forces outside the system are **(6)** ____________________. When the net external force on a closed system is zero, the system is a(n) **(7)** ____________________. The **(8)** ____________________ states that the momentum of any closed system with no net external forces does not **(9)** ____________________. Using this law, you can connect **(10)** ____________________ before and after an interaction without knowing any of the details of the **(11)** ____________________.

In your textbook, read about explosions.

Read the paragraph below. Then circle the letter of the choice that best completes each statement.

Two in-line skaters are skating on such a smooth surface that there are no external forces. They both start from rest. Skater A gives skater B a push in the *positive* direction.

12. The momentum of the system before the push is ____.

a. >0 **b.** 0 **c.** <0

13. The momentum of the system after the push is ____.

a. >0 **b.** 0 **c.** <0

14. The momentum of skater B after the push, p_{B2}, is ____.

a. >0 **b.** 0 **c.** <0

9 Study Guide

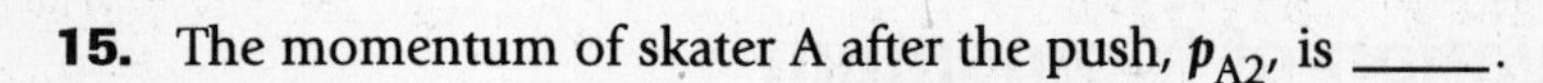

15. The momentum of skater A after the push, p_{A2}, is _____.

a. $+p_{B2}$ **b.** 0 **c.** $-p_{B2}$

16. If the velocity of skater B after the push is represented by v_{B2}, the velocity of skater A after the push, v_{A2}, is _____.

a. $+\left(\frac{m_A}{m_B}\right) v_{B2}$

b. 0

c. $-\left(\frac{m_B}{m_A}\right) v_{B2}$

d. $-\left(\frac{m_A}{m_B}\right) v_{B2}$

In your textbook, read about two-dimensional collisions.

Read the paragraph below. Label the momentum diagram, indicating the final momentum of the system p_2, the momenta of both balls after the collision p_{A2} and p_{B2}, and the 50.0° angle.

A 4.0-kg ball, A, is moving at a speed of 3.0 m/s. It collides with a stationary ball, B, of the same mass. After the collision, ball A moves off in a direction of 50.0° to the left of its original direction. Ball B moves off in a direction of 90.0° to the right of ball A's final direction.

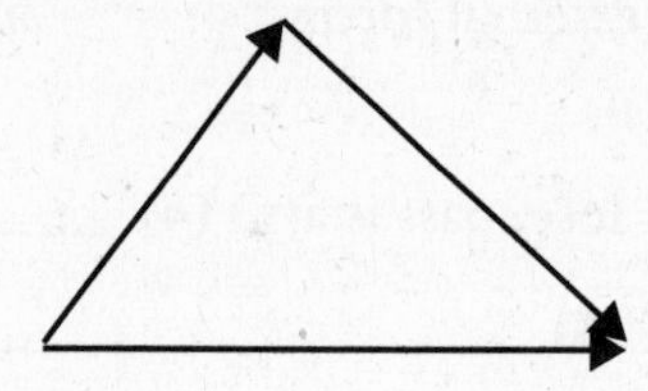

For each situation below, write the letter of the matching item.

_____ **17.** initial momentum of the system

_____ **18.** magnitude of final momentum of the system

_____ **19.** magnitude of ball A's momentum after the collision

_____ **20.** magnitude of ball B's momentum after the collision

a. (12 kg·m/s)(cos 50.0°)

b. 12 kg·m/s

c. (12 kg·m/s)(sin 50.0°)

d. p_A

Date ______________ Period ______________ Name ______________________________

10 Study Guide

Use with Chapter 10.

Energy, Work, and Simple Machines

Vocabulary Review

Write the term that correctly completes each statement. Use each term once.

compound machine	**kinetic energy**	**system**
efficiency	**machine**	**watt**
effort force	**mechanical advantage**	**work**
energy	**power**	**work-energy theorem**
ideal mechanical advantage	**resistance force**	
joule	**simple machine**	

1. ______________________ The energy resulting from motion of an object is called _____.

2. ______________________ The _____ states that the work done on an object is equal to the object's change in kinetic energy.

3. ______________________ The force exerted on a machine is called the _____.

4. ______________________ The ratio of the output work to the input work is a machine's _____.

5. ______________________ The SI unit of work is the _____.

6. ______________________ The unit of power equal to 1 J of energy transferred in 1 s is the _____.

7. ______________________ A machine exerts _____.

8. ______________________ The product of the applied force and the distance through which the force is applied is _____.

9. ______________________ The ability of an object to produce a change in itself or its surroundings is _____.

10. ______________________ The rate of doing work is _____.

11. ______________________ A device that changes the magnitude or direction of a force is a(n) _____.

12. ______________________ The ratio of effort distance to resistance distance is a machine's _____.

13. ______________________ The ratio of resistance force to effort force is a machine's _____.

14. ______________________ A device that consists of two or more simple machines linked so that the resistance force of one machine becomes the effort force of the second machine is a(n) _____.

15. ______________________ A defined group of objects is called a(n) _____.

16. ______________________ A lever, pulley, wheel and axle, inclined plane, wedge, or screw is a(n) _____.

Name ______________________

10 Study Guide

Section 10.1: Energy and Work

In your textbook, read about energy.

For each phrase on the left, write the letter of the matching item.

_____ **1.** symbol for kinetic energy

_____ **2.** calculation of kinetic energy

_____ **3.** symbol for work

_____ **4.** calculation of work

_____ **5.** statement that the work done on an object is equal to the object's change in kinetic energy

_____ **6.** equivalent to 1 $\text{kg}\cdot\frac{\text{m}^2}{\text{s}^2}$

a. W

b. Fd

c. $\frac{mv^2}{2}$

d. $\Delta K = W$

e. 1 J

f. K

In your textbook, read about energy transfer and calculating work.

Circle the letter of the choice that best completes each statement.

7. Through the process of doing work, energy can move between the environment and the system as the result of _____.

a. forces
b. matter
c. momentum
d. impulse

8. If the environment does work on the system, the quantity of work is _____.

a. negative
b. positive
c. zero
d. undetermined

9. If the environment does work on the system, the energy of the system _____.

a. decreases
b. increases
c. remains the same
d. cannot be determined

10. If the system does work on the environment, the energy of the system _____.

a. decreases
b. increases
c. is undetermined
d. remains the same

11. In the equation $W = Fd$, Fd holds only for _____ forces exerted in the direction of displacement.

a. constant
b. frictional
c. gravitational
d. positive

12. In the equation $W = Fd \cos \theta$, angle θ is the angle between F and the _____.

a. x-axis
b. y-axis
c. direction of the displacement
d. vertical direction

Name ____________________

10 Study Guide

For each situation shown, write its letter next to the number of the correct expression.

13. ____________________ $W > 0$

14. ____________________ $W = 0$

15. ____________________ $W < 0$

A

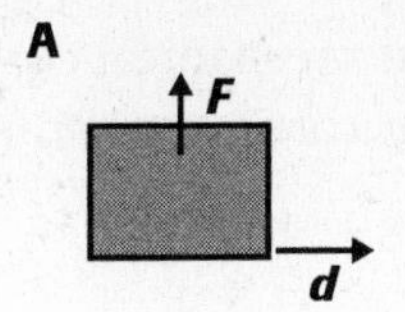

B

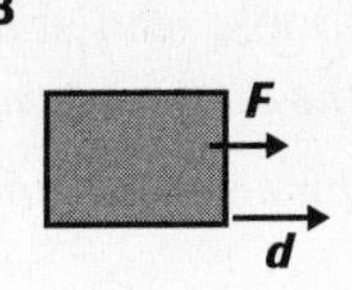

C

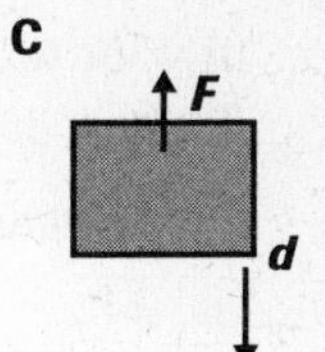

D

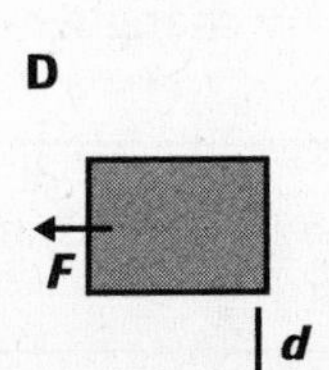

E

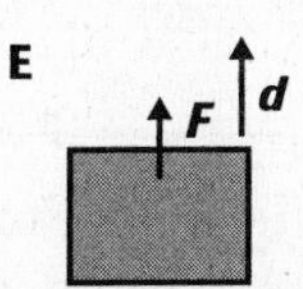

F

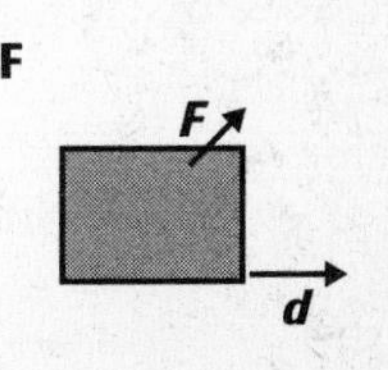

The diagram below is the force-displacement graph of a crate that was pushed horizontally. Answer the following questions.

16. What was the magnitude of the force acting on the crate?

____

17. How far did the crate move horizontally?

____

18. What does the area under the curve of this graph represent?

____

____

19. How much work was done in moving the crate 0.1 m?

____

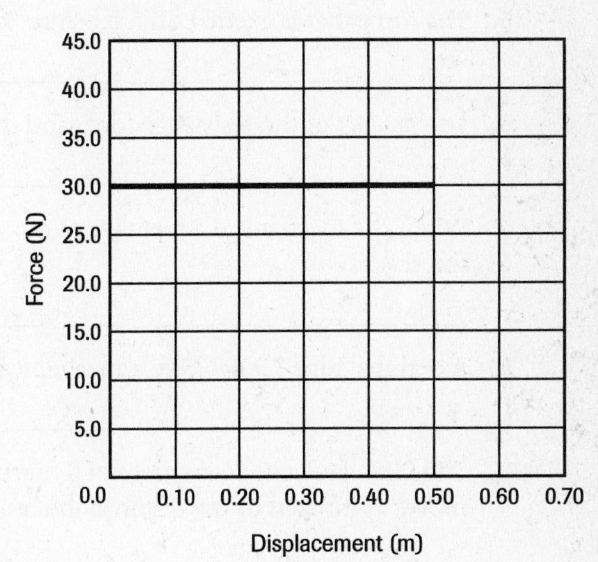

For each term on the left, write the letter of the matching item.

_____ **20.** rate of doing work

_____ **21.** unit of power

_____ **22.** symbol for power

_____ **23.** calculation of power

_____ **24.** 1000 watts

a. power

b. $\frac{W}{t}$

c. watt

d. kW

e. *P*

Name ______________________

10 Study Guide

Section 10.2: Machines

In your textbook, read about mechanical advantage and efficiency.

For each of the statements below, write true *or* rewrite *the italicized part to make the statement true.*

1. A *machine* makes doing a task easier.

____

2. A machine eases the load by changing either the magnitude or the direction of *energy*.

____

3. Work is the transfer of *energy* by mechanical means.

____

4. A machine *cannot* create energy.

____

5. The force that is exerted on a machine is the *effort force*.

____

6. The *product* of the resistance force and the effort force is the mechanical advantage of the machine.

____

7. When the mechanical advantage of a machine is greater than one, the machine *decreases* the effort force.

____

8. A real machine *cannot* have a mechanical advantage that is less than one.

____

9. The *ideal mechanical advantage* of a machine can be used to calculate the distance the effort force moves compared to the distance the resistance force moves.

____

10. If a machine transfers all of the energy applied to it, then the output work is *less than* the input work.

____

11. The *IMA* of most machines is fixed by their *design*.

____

12. The efficiency of a machine is the *ratio* of work output to work input.

____

13. An ideal machine has an efficiency *greater than* 100 percent.

____

Name ______________________

10 Study Guide

14. A real machine has an efficiency *equal to* 100 percent.

____

15. The *lower* the efficiency of a machine, the greater the effort force needed to produce the same resistance force.

____

In your textbook, read about simple machines.

The numbers below correspond to the numbers in the diagram. For each number, write the letter of the matching term from the right column. A term may be used more than once.

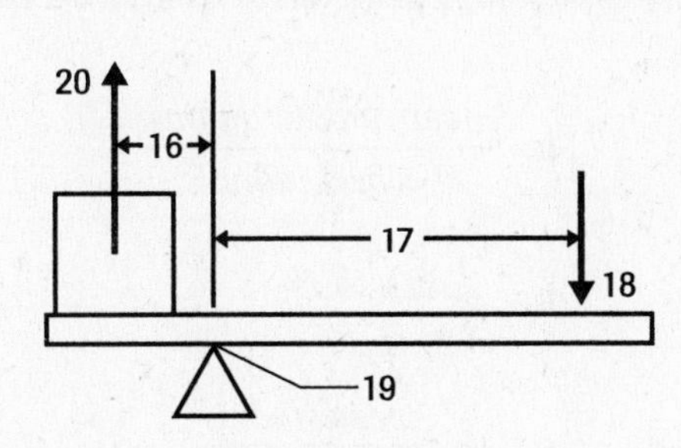

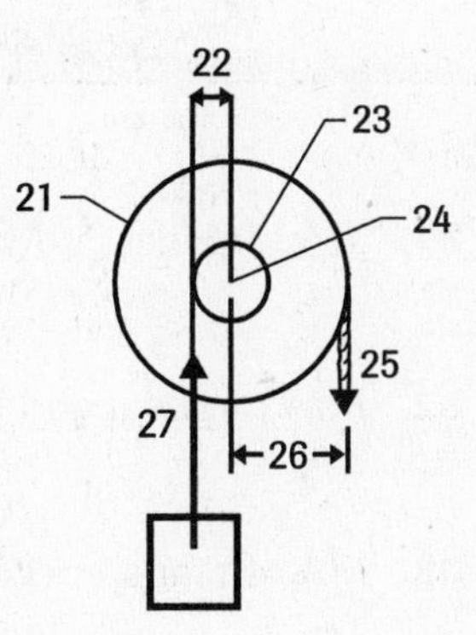

16.	_____	**22.**	_____	**a.**	axle
17.	_____	**23.**	_____	**b.**	F_e
18.	_____	**24.**	_____	**c.**	F_r
19.	_____	**25.**	_____	**d.**	pivot point
20.	_____	**26.**	_____	**e.**	r_e
21.	_____	**27.**	_____	**f.**	r_r
				g.	wheel

Refer to the machine on the left in the diagram above. Circle the letter of the choice that best completes the statement or answers the question.

28. What is the value of the *MA*?

a. >1
b. =1
c. <1
d. cannot be determined

29. What is the value of the *IMA*?

a. >1
b. =1
c. <1
d. cannot be determined

30. To increase the *IMA* of the machine, you would _____.

a. increase r_r
b. increase r_e
c. decrease r_e
d. decrease F_r

10 Study Guide

Name ______________________

In your textbook, read about compound machines.

Refer to the diagram below. For each term on the left, write the letter of the matching item.

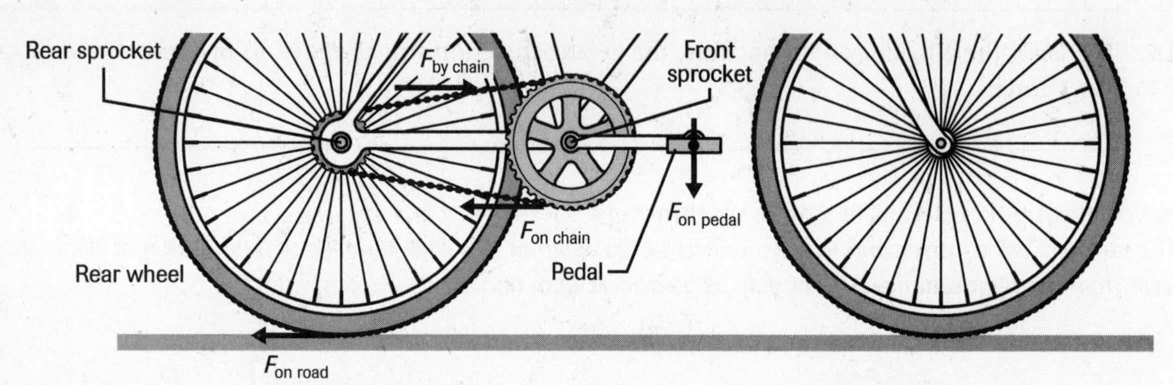

_____ **31.** *MA* of pedal and front sprocket

_____ **32.** *IMA* of pedal and front sprocket

_____ **33.** *MA* of rear sprocket and rear wheel

_____ **34.** *IMA* of rear sprocket and rear wheel

_____ **35.** *MA* of pedal and rear wheel

_____ **36.** *IMA* of pedal and rear wheel

a. $\frac{\text{rear sprocket radius}}{\text{wheel radius}}$

b. $\frac{F_{on\ road}}{F_{by\ chain}}$

c. $\frac{\text{pedal radius}}{\text{front sprocket radius}}$

d. $\frac{F_{on\ road}}{F_{on\ pedal}}$

e. $\frac{\text{pedal radius}}{\text{wheel radius}}$

f. $\frac{\text{rear sprocket radius}}{\text{front sprocket radius}} \times \frac{\text{pedal radius}}{\text{wheel radius}}$

g. $\frac{F_{on\ chain}}{F_{on\ pedal}}$

Refer to the diagram above. Write the value <1, $=1$, or >1 to complete the table below.

Table 1

Machine	*MA* Value	*IMA* Value
pedal and front sprocket	**37.**	**38.**
rear sprocket and rear wheel	**39.**	**40.**
pedal and rear wheel	**41.**	**42.**

Date ______________ Period ______________ Name ______________________

11 Study Guide

Use with Chapter 11.

Energy

Vocabulary Review

Write the term that correctly completes each statement. Use each term once.

closed system	**inelastic collision**	**mechanical energy**
elastic collision	**isolated system**	**momentum**
elastic potential energy	**joule**	**reference level**
energy	**kinetic energy**	**work**
gravitational potential energy	**law of conservation of energy**	**work-energy theorem**

1. momentum — The energy of an object resulting from motion is _____.

2. work-energy theorem — The _____ states that the work done on an object is equal to the objectís change in kinetic energy.

3. elastic potential energy — The energy in a compressed spring is _____.

4. closed system — A system that does not gain or lose mass is a(n) _____.

5. inelastic collision — A collision in which the kinetic energy doesn't change is a(n) _____.

6. joule — The SI unit of work is the _____.

7. elastic collision — A collision in which kinetic energy decreases is a(n) _____.

8. gravitational potential energy — Energy stored in the Earth-object system as a result of gravitational interaction between the object and Earth is _____.

9. work — The product of the applied force and the distance through which the force is applied is _____.

10. energy — The ability of an object to produce a change in itself or its surroundings is _____.

11. law of conservation of energy — Within a closed, isolated system, energy can change form, but the total amount of energy is constant; this is a statement of the _____.

12. mechanical energy — The sum of the kinetic energy and gravitational potential energy is _____.

13. isolated system — A closed system in which the net external force on the system is zero is a(n) _____.

14. reference point — The position at which the potential energy is defined to be zero is a(n) _____.

15. kinetic energy — The product of the mass and velocity of an object is its _____.

Name ______________________

11 Study Guide

Section 11.1: The Many Forms of Energy

In your textbook, read about modeling the work-energy theorem.

The diagram below shows changes to your finances for the first week in October. Circle the choice that best completes the statement or answers the question.

Starting amount
Wages
Birthday gift
Cost of CD
Cost of lunch
Cost of Shirts
Dollars
50
40
30
20
10
0
−10
−20
−30
Mon 10/1
Tues 10/2
Wed 10/3
Thurs 10/4
Fri 10/5
Sat 10/6
Sun 10/7
Date

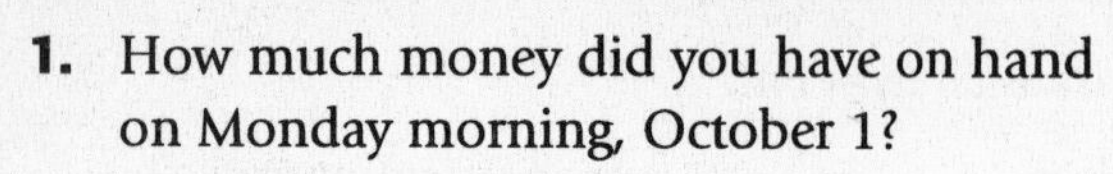

1. How much money did you have on hand on Monday morning, October 1?

   a. $85 c. $40
   b. $45 d. $0

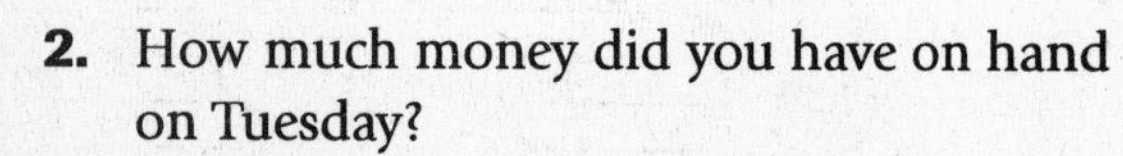

2. How much money did you have on hand on Tuesday?

   a. $85 c. $40
   b. $45 d. $0

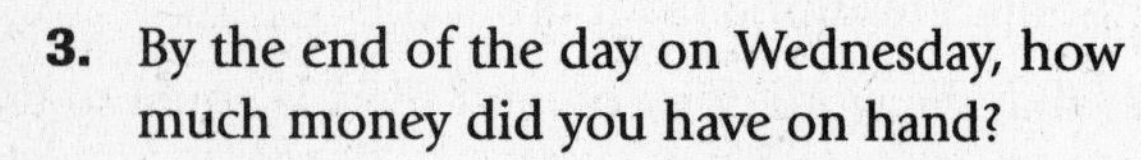

3. By the end of the day on Wednesday, how much money did you have on hand?

   a. $85 b. $45 c. $40 d. $0

4. On Thursday, what happened to the amount of money you had on hand?

   a. It increased. b. It decreased. c. It remained unchanged.

5. By the end of the day on Saturday, how much money had you spent during the week?

   a. $75 b. $50 c. $40 d. $0

6. How much money did you have on hand on the morning of Monday, October 8?

   a. $75 b. $55 c. $40 d. $0

Read the following situations and draw before-and-after energy diagrams in the space provided.

7. A partially loaded shopping cart has a kinetic energy of 60 J. A short time later the cart's kinetic energy is 60 J.

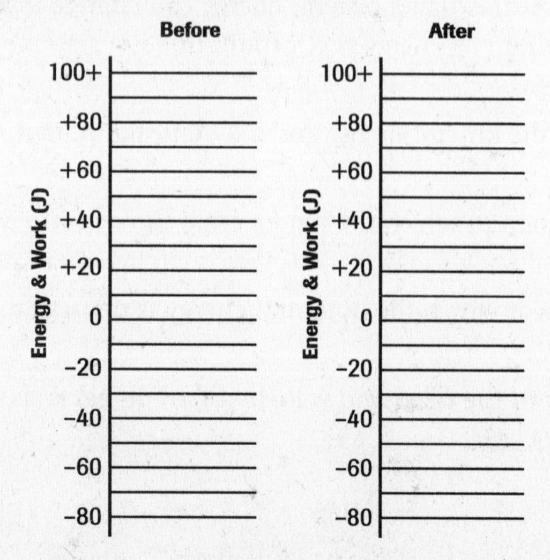

8. A partially loaded shopping cart has a kinetic energy of 60 J. You do 20 J of work on the cart.

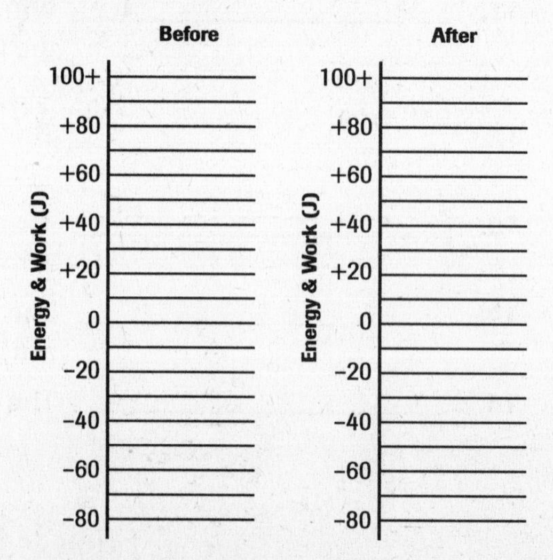

Name ______________________________

11 Study Guide

9. You do 50 J of work on a shopping cart at rest.

Before

After

Energy & Work (J)

100+ +80 +60 +40 +20 0 -20 -40 -60 -80

10. A partially loaded shopping cart has a kinetic energy of 60 J. You do −40 J of work on the cart.

Before

After

Energy & Work (J)

100+ +80 +60 +40 +20 0 -20 -40 -60 -80

11. You do 70 J of work on a partially loaded shopping cart, giving it a total kinetic energy of 80 J.

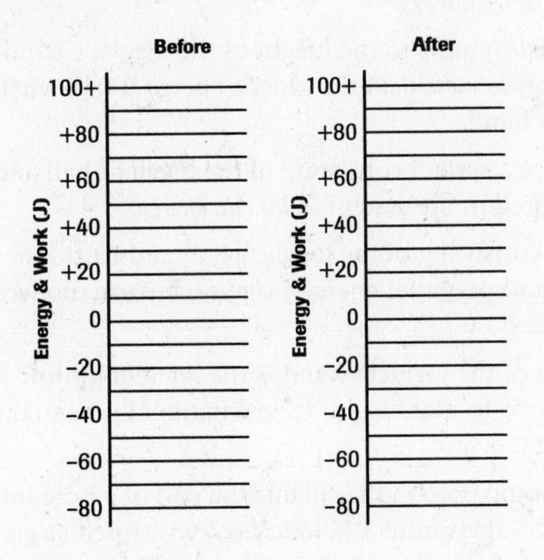

12. A partially filled shopping cart has a kinetic energy of 80 J. The work you do on the cart stops it.

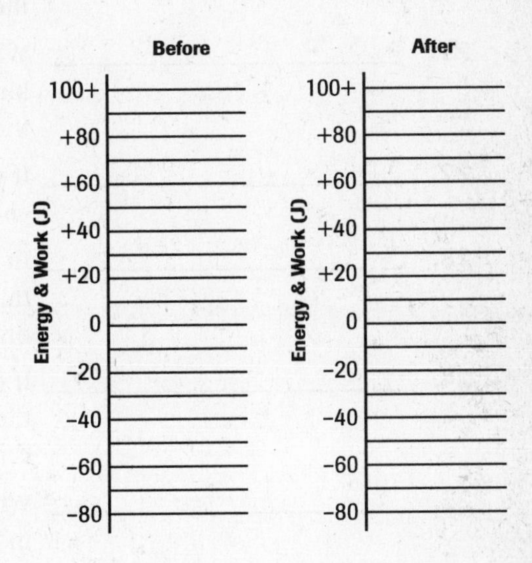

Name ______________________________

11 Study Guide

In your textbook, read about kinetic energy.

Circle the letter of the choice that best answers each question. Refer to Table 11-1 on page 250 of your textbook.

13. What accounts for the aircraft carrier having a greater kinetic energy than the orbiting satellite?

a. mass **b.** speed **c.** mass and speed

14. What accounts for the pitched baseball having a greater kinetic energy than the falling nickel?

a. mass **b.** speed **c.** mass and speed

15. What accounts for the bumblebee having more kinetic energy than the snail?

a. mass **b.** speed **c.** mass and speed

16. How much work would have to be done to the truck starting from rest to give it half the listed kinetic energy?

a. 0 J **b.** 2.2×10^3 J **c.** 5.7×10^5 J **d.** 1.1×10^6 J

17. How fast would the truck be moving if it had half the listed kinetic energy?

a. –52 km/h **b.** 0 km/h **c.** 71 km/h **d.** 118 km/h

In your textbook, read about gravitational and elastic potential energies.

For each statement below, write true *or rewrite the italicized part to make the statement true.*

18. ______________________ If a juggleris ball is considered a system, work is done on it by gravity and *the juggler's hand.*

19. ______________________ On the way up, the work done on the ball by gravity is $-mgh$.

20. ______________________ On the way down, the work done on the ball by gravity *decreases* the ball's kinetic energy.

21. ______________________ When the ball returns to the height of the juggler's hand, its kinetic energy is *more than* the kinetic energy it had when it left the juggler's hand.

22. ______________________ If you choose a system consisting of the juggler's ball and Earth, energy is stored in the system as *kinetic energy.*

23. ______________________ In a system consisting of the juggler's ball and Earth, the sum of the kinetic and potential energies *changes* because no work is done on the system.

24. ______________________ If the height of the juggler's hand is the reference point, when the ball returns to that height, its gravitational potential energy is *less than* 0 J.

25. ______________________ When a running pole-vaulter plants the end of a pole into a socket in the ground, the vaulter's kinetic energy is stored as *gravitational potential energy* in the bent pole.

Name ____________________

11 Study Guide

Section 11.2: Conservation of Energy

In your textbook, read about the conservation of mechanical energy.

The diagram below shows the position of two identical balls initially at rest. Complete the following table, assuming there is neither air resistance nor frictional forces on either ball. Using the table as a guide, construct graphs of E, U_g, and K versus d for ball A on the grid. Then answer the following question.

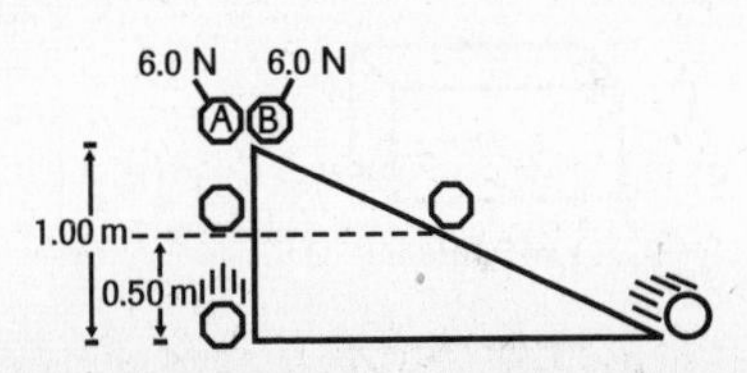

Table 1

	Ball A			Ball B		
Location (*d*)	***E***	U_g	***K***	***E***	U_g	***K***
1.00 m						
0.50 m						
0.00 m						

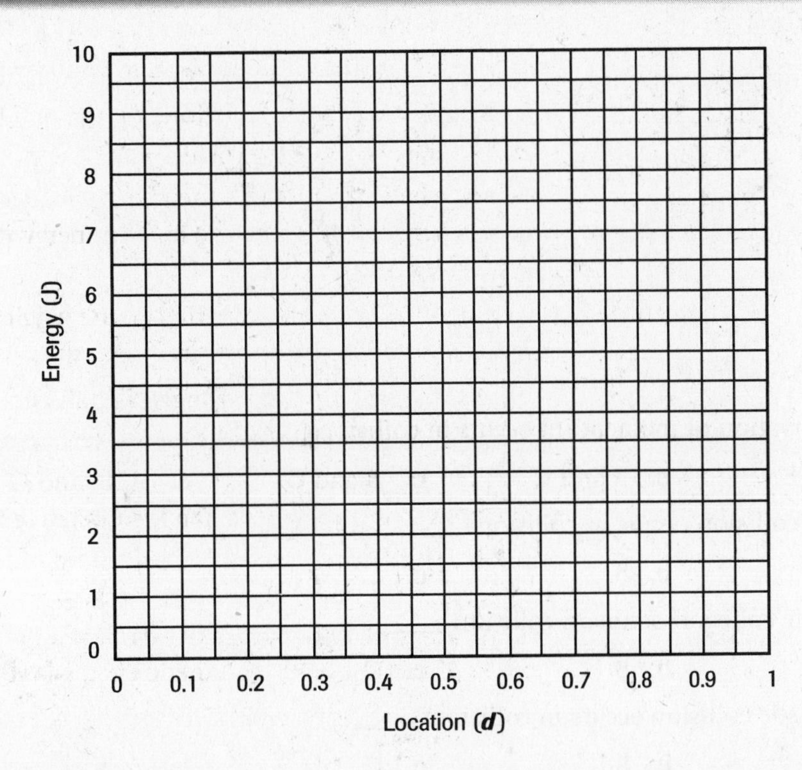

1. How would the graphs of E, U_g, and K versus d for ball B compare with the graphs you constructed for ball A?

____

Name ______________________

11 Study Guide

In your textbook, read about analyzing collisions.

The diagrams below show the motion of two identical 2.0 kg boxes before and after they collide. Sketch bar graphs of the energy of each system before and after the collision. Circle the letter of the choice that best completes each statement.

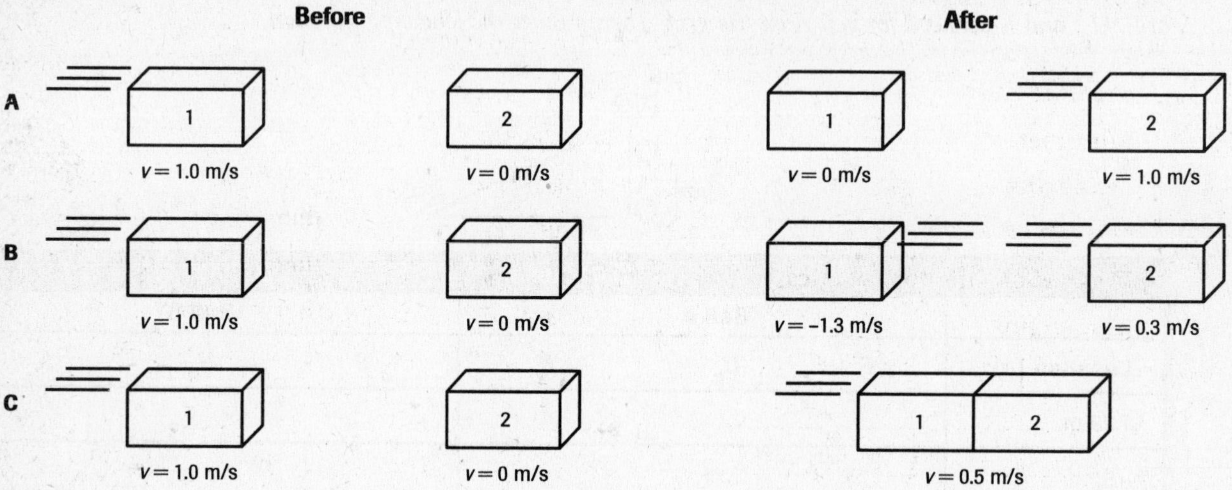

Before **After**

A
1 $v = 1.0$ m/s 2 $v = 0$ m/s 1 $v = 0$ m/s 2 $v = 1.0$ m/s

B
1 $v = 1.0$ m/s 2 $v = 0$ m/s 1 $v = -1.3$ m/s 2 $v = 0.3$ m/s

C
1 $v = 1.0$ m/s 2 $v = 0$ m/s 1 2 $v = 0.5$ m/s

2. The conservation of momentum occurs in collisions _____.

a. A and B **b.** A and C **c.** B and C **d.** A, B, and C

3. An elastic collision occurs in collision _____.

a. A **b.** B **c.** C

4. An inelastic collision occurs in collision _____.

a. A **b.** B **c.** C

5. A superelastic collision occurs in collision _____.

a. A **b.** B **c.** C

6. The conservation of energy occurs in collisions _____.

a. A and B **b.** A and C **c.** B and C **d.** A, B, and C

Date ______________ Period ______________ Name ______________________

12 Study Guide

Use with Chapter 12.

Thermal Energy

Vocabulary Review

Write the term that correctly completes each statement. Use each term once.

absolute zero	**heat of fusion**	**second law of thermodynamics**
calorimeter	**heat of vaporization**	**specific heat**
conduction	**kelvin melting point**	**thermal energy**
entropy	**kinetic-molecular theory**	**thermodynamics**
first law of thermodynamics	**melting point**	**thermometer**
heat engine	**radiation**	

1. ______________________ A device that measures temperature is a(n) _____.

2. ______________________ The transfer of kinetic energy when particles collide is _____.

3. ______________________ The _____ states that natural processes go in a direction that maintains or increases the universe's total entropy.

4. ______________________ The _____ states that matter is made up of particles that are always in motion.

5. ______________________ A device that continuously converts thermal energy to mechanical energy is a(n) _____.

6. ______________________ The temperature at which a substance changes from solid to liquid is called the _____.

7. ______________________ The lowest possible temperature is _____.

8. ______________________ A device that measures changes in thermal energy is a(n) _____.

9. ______________________ The transfer of energy by electromagnetic waves is _____.

10. ______________________ The amount of energy that must be added to raise the temperature of a unit mass one unit is the _____ of a material.

11. ______________________ The unit of temperature on the Kelvin scale is the _____.

12. ______________________ The amount of energy needed to melt 1 kg of a substance is the _____ of the substance.

13. ______________________ The _____ states that the total increase in thermal energy of a system is the sum of the work done on it and the heat added to it.

14. ______________________ The disorder in a system is _____.

15. ______________________ The study of heat is _____.

16. ______________________ The overall energy of motion of the particles that make up the object is the _____ of the object.

17. ______________________ The amount of thermal energy needed to change 1 kg of liquid to a gas is the _____.

Name ________________________

12 Study Guide

Section 12.1: Temperature and Thermal Energy

In your textbook, read about the theory and measurement of thermal energy.

For each of the statements below, write true or rewrite the italicized part to make the statement true.

1. ____________________ In a hot body, particles tend to move *faster* than in a cooler body.

2. ____________________ At a given temperature, all the particles in a body have *the same* energy.

3. ____________________ The thermal energy in an object *doesn't* depend on the number of particles in it.

4. ____________________ Two bodies are in *thermal* equilibrium if they are at the same temperature.

5. ____________________ A kelvin is equal in size to a *Celsius* degree.

6. ____________________ Absolute zero is equal to *0°C.*

7. ____________________ The equation that allows conversion of a Celsius temperature to a Kelvin temperature is $T_k + 273.15 = T_c$.

8. ____________________ The *caloric* theory could successfully explain why objects become warm when they are rubbed together.

In your textbook, read about temperature measurement.

Express each temperature in the scale indicated.

9. The freezing point of water, in degrees Celsius ____________________

10. The freezing point of water, in kelvins ____________________

11. The boiling point of water, in degrees Celsius ____________________

12. The boiling point of water, in kelvins ____________________

13. The lowest possible temperature, in degrees Celsius ____________________

14. The lowest possible temperature, in kelvins ____________________

In your textbook, read about thermal-energy transfer.

For each term on the left, write the corresponding symbol.

_____	**15.** heat	C
_____	**16.** degree Celsius	ΔT
_____	**17.** kelvin	°C
_____	**18.** specific heat	Q
_____	**19.** energy change	K
_____	**20.** temperature change	ΔE

Name ______________________

12 Study Guide

In your textbook, read about thermal-energy transfer and specific heat.

Answer the following questions.

21. Describe the typical direction of heat transfer when a cooler body is in contact with a warmer one.

____

22. Write the equation for calculating the amount of heat transferred, given the mass, specific heat, and temperature change.

____

23. Name the three means of thermal transfer.

____

____

24. How does the specific heat of liquid water compare in general with that of most other substances?

____

____

In your textbook, read about calorimetry.

The diagram below shows a calorimeter just after addition of an object and after thermal equilibrium is reached. Refer to the diagram to answer the following questions.

Initial
Final
Calorimeter
Water
Object

25. What is the approximate value of T_f?

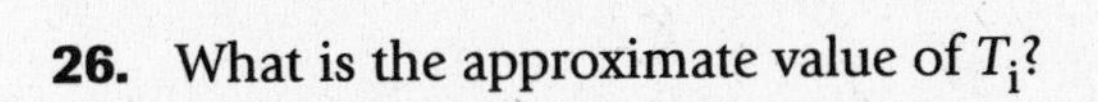

26. What is the approximate value of T_i?

____

27. Is the value of ΔT for the water positive or negative?

____

28. Has heat flowed from the object to the water, or to the object from the water?

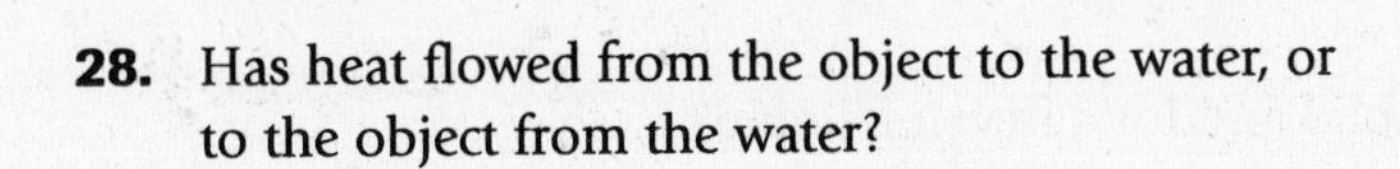

29. Is Q for the object positive or negative?

____

30. Is Q for the water positive or negative?

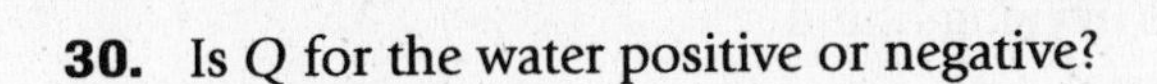

31. Assuming no heat loss to the surroundings, how do the ΔE values for the water and the object compare?

____

Name ____________________

12 Study Guide

Section 12.2: Change of State and Laws of Thermodynamics

In your textbook, read about changes of state.

Circle the letter of the choice that best completes the statement or answers the question.

1. The energy needed to melt 1 kg of a substance is called the _____.

a. heat of vaporization

b. entropy

c. heat of fusion

d. melting point

2. The temperature at which all added thermal energy converts a material from liquid to gas is the _____.

a. melting point

b. boiling point

c. heat of fusion

d. heat of vaporization

3. How do the specific heats of ice, liquid water, and steam compare?

a. They are equal.

b. That of ice is highest.

c. That of steam is highest.

d. That of liquid water is highest.

4. Which equation correctly relates heat, mass, and heat of vaporization?

a. $Q = m/H_V$

b. $Q = H_V/m$

c. $Q = m + H_V$

d. $Q = mH_V$

5. Is Q positive or negative for a melting solid and for a freezing liquid?

a. negative in both cases

b. positive in both cases

c. positive for the melting solid, negative for the freezing liquid

d. negative for the melting solid, positive for the freezing liquid

Name ______________________

12 Study Guide

In your textbook, read about changes of state.

For each of the statements below, write true *or rewrite the italicized part to make it true.*

6. ______________________ When a liquid changes to a gas, it *releases* thermal energy.

7. ______________________ During melting, the kinetic energy of the particles *does not* change.

8. ______________________ An increase in the range of particle speeds in an object represents an *increase* in entropy.

9. ______________________ The heat required to melt a solid *does not* depend on the mass of the sample.

10. ______________________ In a plot of temperature versus heat added, the portion of the graph representing the boiling process is *horizontal.*

In your textbook, read about the first and second laws of thermodynamics.

Decide whether each statement relates more closely to the first law of thermodynamics or to the second law. Write first law *or* second law.

11. Total entropy tends to increase.

____

12. Energy is neither created nor destroyed.

____

13. Thermal-energy increases depend on the work done and the heat added.

____

14. Heat does not flow from a cold body to a hot body.

____

15. Waste heat is generated whenever thermal energy is converted into mechanical energy.

____

16. Energy can be converted from one form to another.

____

17. The total amount of useful energy tends to decrease.

____

Name ______________________

12 Study Guide

In your textbook, read about devices that convert energy.

For each of the statements below, write true *or rewrite the italicized part to make it true.*

18. A *heat engine* converts thermal energy to mechanical energy.

____

19. A heat engine requires a low-temperature receptacle into which *mechanical* energy can be delivered.

____

20. An automobile internal combustion engine is an example of a *heat pump.*

____

21. Using mechanical energy, a refrigerator removes thermal energy from a warmer body.

____

22. A heat pump is a refrigerator that can run in *one direction.*

____

In your textbook, read about entropy.

Decide whether the process is likely to occur spontaneously. Write yes *or* no.

23. A hot body absorbs thermal energy from a cooler one. ____________

24. A set of particles becomes more orderly. ____________

25. An engine generates waste heat. ____________

26. Gas particles collect in one corner of a container. ____________

27. Total entropy increases. ____________

28. Usable energy increases. ____________

Date ______________ Period ______________ Name ______________________

13 Study Guide

Use with Chapter 13.

States of Matter

Vocabulary Review

Write the term that correctly completes each statement. Use each term once.

adhesion	capillary action	evaporation	pressure
amorphous solid	cohesive force	fluid	surface tension
Archimedes' principle	condensation	pascal	thermal expansion
Bernoulli's principle	crystal lattice	Pascal's principle	volatile
buoyant force	elasticity	plasma	

1. ______________ The SI unit of pressure is the _____.
2. ______________ An upward force on an object in a liquid is a(n) _____.
3. ______________ A _____ contains electrons and positively charged ions.
4. ______________ A fixed pattern of particles within a solid is a(n) _____.
5. ______________ An increase in temperature may cause _____.
6. ______________ "The magnitude of the buoyant force on an object equals the weight of fluid displaced by the object" is a statement of _____.
7. ______________ "A change in pressure at any point on a confined fluid is transmitted undiminished throughout the fluid" is a statement of _____.
8. ______________ The tendency of the surface of a liquid to contract to the smallest possible area is _____.
9. ______________ A liquid that evaporates quickly is _____.
10. ______________ The ability of an object to return to its original form when external forces are removed is _____.
11. ______________ "As the velocity of a fluid increases, the pressure exerted by the fluid decreases" is a statement of _____.
12. ______________ The escape of particles from a liquid's surface is _____.
13. ______________ Force divided by area equals _____.
14. ______________ The attractive force between particles of different substances is _____.
15. ______________ A substance that has definite volume and shape but no regular crystal structure is a(n) _____.
16. ______________ The rise of a liquid up a narrow tube is due to _____.
17. ______________ The change of a gas to a liquid is _____.
18. ______________ An attractive force between particles of the same substance, caused by electromagnetic attraction, is a(n) _____.
19. ______________ A _____ flows and has no definite shape of its own.

Name ____________________

13 Study Guide

Section 13.1: The Fluid States

In your textbook, read about measures and equivalents for pressure.

For each term on the left, write the corresponding symbol.

________	**1.** symbol for pressure	$1\ N/m^2$
________	**2.** symbol for weight	101 kPa
________	**3.** symbol for surface area	A
________	**4.** unit of pressure	F/A
________	**5.** unit of weight	F_g
________	**6.** definition of pressure	N
________	**7.** equivalent to 1 Pa	P
________	**8.** atmospheric pressure at sea level	Pa

In your textbook, read about pressure and fluids.

For each of the statements below, write true *or* rewrite *the italicized part to make the statement true.*

9. The shape of a container *does not affect* the pressure at any given depth of the fluid it contains.

____

10. The buoyant force on an object depends on the weight of the *object.*

____

11. The *apparent weight* of an object in water is equal to its actual weight minus the upward buoyant force.

____

12. Fluids flow and have no definite *volume.*

____

13. The pressure of the water on a body beneath its surface equals ρhg.

____

14. Bernoulli's principle applies to *turbulent* flow only.

____

Name ______________________

13 Study Guide

In your textbook, read about Bernoulli's principle, Archimedes' principle, and Pascal's principle.

Complete the table below by writing the name of the principle that is most clearly illustrated in each situation. Choose among Bernoulli's principle, Archimedes' principle, and Pascal's principle.

Situation	Principle Illustrated
15. A man uses the hydraulic brake to stop his truck.	______________
16. An airplane rises into the air.	______________
17. An ice cube floats in water.	______________
18. A piece of paper rises when wind blows across it.	______________
19. A fish uses its air bladder to move downward in water.	______________
20. Ketchup rises up in a squeezed plastic bottle.	______________
21. A person feels lighter in water than on land.	______________
22. Curved airfoils help hold down the rear wheels of race cars.	______________

Write the term that correctly completes each statement.

The relationship between the **(23)** ______________ of a moving fluid and the **(24)** ______________ exerted by the fluid is described by Bernoulli's principle. Airfoils are devices that use this principle to produce **(25)** ______________, a net force in a(n) **(26)** ______________ direction, when moving through a(n) **(27)** ______________. An airplane wing is curved so that the **(28)** ______________ surface has a greater curve than the **(29)** ______________ surface. Air moving over the top of the wing moves a(n) **(30)** ______________ distance, and therefore travels at a greater speed. The air pressure over the wing is **(31)** ______________ than the pressure under the wing. The flow of a fluid can be represented by a set of lines called **(32)** ______________. The closer the lines are to each other, the greater is the **(33)** ______________ and the lower is the **(34)** ______________ of the fluid. Swirling of such lines indicates that the fluid flow is **(35)** ______________, and Bernoulli's principle does not apply to the situation.

Name ______________________

13 Study Guide

In your textbook, read about cohesive and adhesive forces.

The diagram on the left shows water molecules in a glass tube. The diagram on the right shows mercury atoms in a glass tube. The longer the vector arrows are, the greater is the force between the particles. Refer to the diagrams to answer the following questions. Use complete sentences.

36. Which arrows represent cohesive forces?

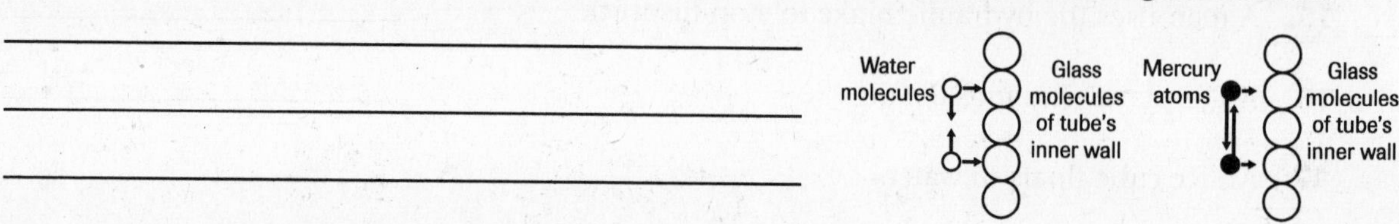

37. Which arrows represent adhesive forces?

38. For the water molecules, how do the cohesive forces compare to the adhesive forces?

39. For the mercury atoms, how do the cohesive forces compare to the adhesive forces?

40. Explain how the behavior of mercury would compare to that of water if a thin glass tube were placed in each liquid.

41. Would the top surface of mercury in a glass container be the same shape as the top surface of water in a similar container? Give the reason for your answer.

Name ______________________

13 Study Guide

Section 13.2: The Solid State

In your textbook, read about solids and thermal expansion of solids.

For each of the statements below, write true *or* false.

________ **1.** The particles in a crystalline solid do not move at all.

________ **2.** Butter and glass are examples of amorphous solids.

________ **3.** Elasticity is the ability of an object to return to its original form when external forces are removed.

________ **4.** Malleability and ductility depend upon elasticity.

________ **5.** Copper is a nonductile metal.

________ **6.** Gold is a malleable metal.

________ **7.** The symbol for the coefficient of linear expansion is α.

________ **8.** Water is most dense at 0°C.

________ **9.** Metals used in a bimetallic strip have different rates of thermal expansion.

In your textbook, read about solids and thermal expansion.

Circle the letter of the choice that best completes the statement or answers the question.

10. For what substances is the solid state less dense than the liquid state?

a. no liquids

b. most liquids other than water

c. water

d. all liquids

11. A substance that can be flattened and shaped into thin sheets by hammering is _____.

a. inelastic

b. adhesive

c. ductile

d. malleable

12. A substance that can be pulled into thin strands of wire is said to be _____.

a. inelastic

b. ductile

c. adhesive

d. malleable

13. The change in volume of a material due to thermal expansion equals _____.

a. $\beta V \Delta T$

b. $\beta V/\Delta T$

c. $\alpha V \Delta T$

d. $\alpha V/\Delta T$

14. The coefficient of linear expansion equals _____.

a. $\Delta T \Delta L$

b. $L_1/\Delta L \Delta T$

c. $\Delta T/\Delta L$

d. $\Delta L/L_1 \Delta T$

Name ______________________

13 Study Guide

In your textbook, read about thermal expansion of solids.
For each term on the left, write the corresponding symbol.

________	**15.** change of length	ΔT
________	**16.** coefficient of volume expansion	α
________	**17.** coefficient of linear expansion	ΔV
________	**18.** change of temperature	$(°C)^{-1}$
________	**19.** change of volume	ΔL
________	**20.** unit for coefficient of linear expansion	β

In your textbook, read about plasma and thermal expansion of liquids.
Use the terms below to complete the passage. You may need to use terms more than once.

electrons	**gas**	**liquid**	**temperature**
fluid	**ions**	**plasma**	

Heating a solid substance eventually turns it to a **(21)** ______________ and then to a **(22)** ______________. Further increases in **(23)** ______________ eventually cause collisions between particles to become violent enough that **(24)** ______________ are pulled off the atoms. The resulting state of matter is called **(25)** ______________. Along with liquids and gases, matter in this state is a **(26)** ______________. It contains positively charged **(27)** ______________ and negatively charged **(28)** ______________. The main difference between a gas and a plasma is that only the **(29)** ______________ can conduct electricity.

Date ______________ Period ______________ Name ______________________

14 Study Guide

Use with Chapter 14.

Waves and Energy Transfer

Vocabulary Review

Write the term that correctly completes each statement. Use each term once.

antinode	destructive interference	node	transverse wave
constructive interference	interference	normal	trough
crest	longitudinal wave	standing wave	wave
	mechanical wave	surface wave	wave pulse

1. ______________________ Any of the low points of a wave is a(n) _____.

2. ______________________ Superposition of waves with opposite amplitudes causes _____.

3. ______________________ Any rhythmic disturbance that carries energy through matter or space is a(n) _____.

4. ______________________ A wave that appears to be standing still is a(n) _____.

5. ______________________ A wave that requires a physical medium is a(n) _____.

6. ______________________ A single bump or disturbance that travels through a medium is a(n) _____.

7. ______________________ Superposition of waves with displacements in the same direction causes _____.

8. ______________________ An unmoving point at which destructively interfering pulses meet is a(n) _____.

9. ______________________ In a(n) _____, the disturbance moves parallel to the direction of motion of the wave.

10. ______________________ A point of largest displacement caused by the constructive interference of waves is a(n) _____.

11. ______________________ Any of the high points of a wave is a(n) _____.

12. ______________________ The result of any superposition of two or more waves is _____.

13. ______________________ In a(n) _____, the particles move in a direction that is both parallel and perpendicular to the direction of wave motion.

14. ______________________ In a(n) _____, the particles vibrate in a direction perpendicular to the direction of motion of the wave.

15. ______________________ A line that is at a right angle to a barrier is a(n) _____.

Name ______________________________

14 Study Guide

Section 14.1: Wave Properties

In your textbook, read about different kinds of waves.
Circle the letter of the choice that best completes each statement.

1. Waves typically transmit _____.

a. matter only
b. both matter and energy
c. energy only
d. neither matter nor energy

2. Water waves, sound waves, and waves that travel down a spring are all examples of _____.

a. mechanical waves
b. transverse waves
c. electromagnetic waves
d. longitudinal waves

3. If a wave moves in a leftward direction but the particles of the medium move up and down, the wave must be a _____.

a. mechanical wave
b. transverse wave
c. surface wave
d. longitudinal wave

4. A surface wave has characteristics of _____.

a. longitudinal waves only
b. transverse waves only
c. both longitudinal and transverse waves
d. neither longitudinal nor transverse waves

For each of the statements below, write true *or rewrite the italicized part to make the statement true.*

5. Sound is an example of a *transverse* wave.

____

6. Fluids usually transmit only *longitudinal* waves.

____

7. Waves at the top of a lake or an ocean are examples of *surface* waves.

____

8. Energy from the sun reaches Earth by means of *longitudinal* electromagnetic waves.

____

9. Waves that travel down a rope are examples of *surface* waves.

____

Name ______________________________

14 Study Guide

In your textbook, read about amplitude and frequency.

For each of the statements below, write true *or rewrite the italicized part to make it true.*

10. *One hertz* is equal to one oscillation per second.

____

11. The speed of a wave generally depends only on the *source.*

____

12. Waves with larger amplitudes tend to transmit *less* energy than do waves with smaller amplitudes.

____

13. The frequency of a wave does not depend on the *medium.*

____

14. A wave of *high* frequency also has a long period.

____

15. The higher the frequency of a wave is, the *longer* the period is.

____

In your textbook, read about measuring waves.

For each term on the left, write the matching symbol from the list on the right.

_____ **16.** wavelength — Δd

_____ **17.** period — v

_____ **18.** speed — T

_____ **19.** frequency — Δt

_____ **20.** displacement of wave peak — f

_____ **21.** time interval — λ

In your textbook, read about the relationship of speed to wave attributes.

Use symbols to write the equation that combines the attributes on the left.

22. frequency and period $f =$ ______________________________

23. speed, displacement, and time interval $v =$ ______________________________

24. speed, wavelength, and frequency $v =$ ______________________________

25. speed, wavelength, and period $v =$ ______________________________

Name ______________________

14 Study Guide

In your textbook, read about measuring waves.
Write the terms that correctly completes the passage.

The **(26)** ______________________ of a wave is its maximum displacement from its position of rest. Doubling the amplitude of a wave increases the rate of energy transfer by a factor of **(27)** ______________________. The time it takes for the motion of an oscillator to repeat itself in generating a wave is the **(28)** ______________________ of the wave. Because the motion keeps repeating, such a wave is a(n) **(29)** ______________________ wave. The number of complete oscillations per second is the **(30)** ______________________ and is measured in units called **(31)** ______________________. Both period and frequency of a wave depend only on its **(32)** ______________________. A material that carries the energy of a mechanical wave is a(n) **(33)** ______________________.

In your textbook, read about the attributes of a wave.
Refer to the drawing below to answer the following questions.

34. Which of the lettered distances represents the wavelength of wave 1?

____

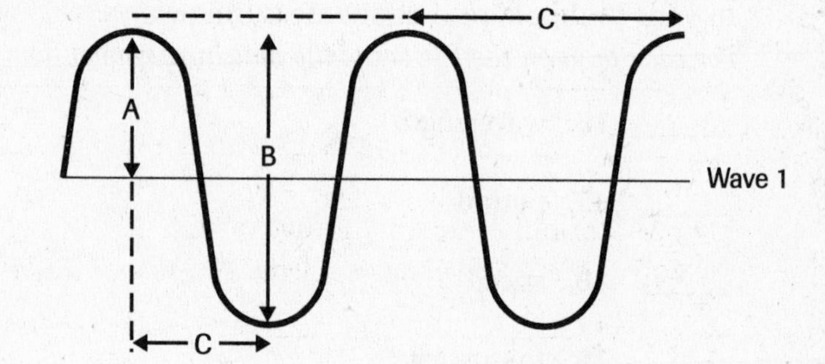

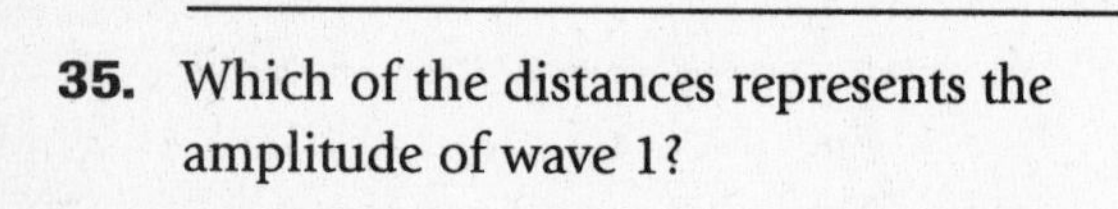

35. Which of the distances represents the amplitude of wave 1?

____

36. Which of the two waves has the greater wavelength?

____

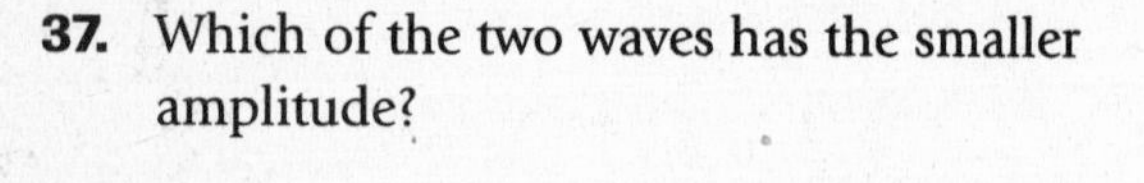

37. Which of the two waves has the smaller amplitude?

____

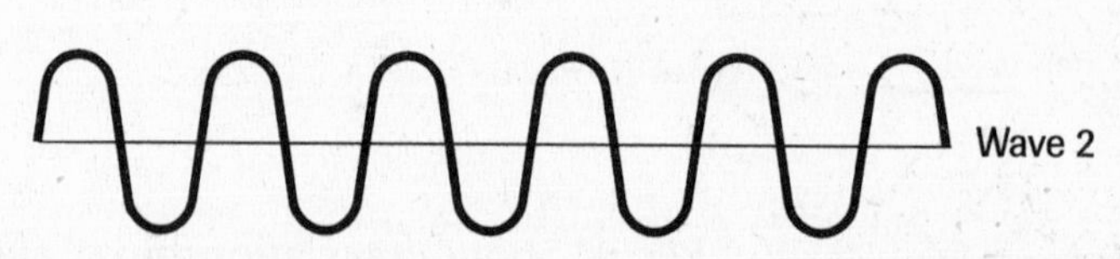

38. Assuming that the two waves move at equal speed, which has the higher frequency?

____

39. Assuming that the two waves move at equal speed, which has the greater period?

____

Name ________________________

14 Study Guide

Section 14.2: Wave Behavior

In your textbook, read about wave reactions to boundaries.

For each of the statements below, write true *or rewrite the italicized part to make it true.*

1. ________________ When a wave encounters the boundary of the medium in which it is traveling, the wave *always* reflects completely back into the medium.

2. ________________ A wave that strikes a boundary is called the *incident* wave.

3. ________________ If a wave passes from a lighter, more flexible spring to a heavier, stiffer one, the resulting reflected wave will be *inverted.*

4. ________________ If a wave pulse is sent down a spring connected to a rigid wall, *almost none* of the wave's energy is reflected back.

Answer the following questions, using complete sentences.

5. What happens to the wavelength of water waves as they move from deep to shallow water?

____

6. What happens to the frequency of water waves as they move from deep to shallow water?

____

7. What happens to the speed of water waves as they move from deep to shallow water? Describe an equation that would show this change.

____

____

In your textbook, read about wave interference.

For each of the statements below, write true *or rewrite the italicized part to make it true.*

8. ________________ According to the principle of superposition, the medium's displacement caused by two or more waves is the algebraic *difference between* the displacements caused by the individual waves.

9. ________________ In constructive interference, the resulting wave has an amplitude *smaller* than that of any of the individual waves.

10. ________________ Two waves *cannot* exist in the same medium at the same time.

11. ________________ A *continuous* wave appears to be standing still.

12. ________________ If the trough of one wave occurs at the same point as the *crest* of another wave, the two may cancel out, producing zero amplitude at that point.

13. ________________ When a continuous wave crosses a boundary and moves more *slowly,* the amplitude increases.

Name ______________________________

14 Study Guide

In your textbook, read about two-dimensional waves.
Answer the following questions, using complete sentences.

14. What is a ray?

____

15. What is a ripple tank?

____

____

16. State the law of reflection, and explain how each of the two kinds of angles involved are measured.

____

____

____

____

17. What is meant by refraction?

____

18. In how many dimensions do the waves travel in the following three cases: a wave moving along a rope, a wave on the surface of water, and a sound wave.

____

____

____

In your textbook, read about diffraction and interference of waves.
Write the terms that correctly complete the passage.

When waves reach a small opening in a barrier, they form waves that have a(n) **(19)** ____________________ shape. This effect is an example of **(20)** ____________________. The smaller the **(21)** ____________________ is, compared to the size of the barrier, the **(22)** ____________________ pronounced this effect is. When there are two closely spaced openings in a barrier, two sets of **(23)** ____________________ waves are produced. Nodes lie along lines between lines of **(24)** ____________________. Nodes form where there is **(25)** ____________________ interference.

Date ______________ Period ______________ Name ______________________

15 Study Guide

Use with Chapter 15.

Sound

Vocabulary Review

Write the term that correctly completes each statement. Use each term once.

beat	dissonance	harmonic	sound level
consonance	Doppler shift	octave	timbre
decibel	fundamental	pitch	

1. ______________ The amplitude of sound, as measured on a logarithmic scale, is called ____.
2. ______________ Any higher frequency that is a multiple of the fundamental frequency is a(n) ____.
3. ______________ An oscillation of wave amplitude caused by the sounding of two nearly identical frequencies is called a(n) ____.
4. ______________ A unit used to measure sound level is the ____.
5. ______________ A jarring combination of pitches is called ____.
6. ______________ Two frequencies in a 1:2 ratio differ by a(n) ____.
7. ______________ The lowest frequency that will resonate in a pipe is the ____.
8. ______________ The apparent change in pitch caused by the motion of a sound source is called the ____.
9. ______________ The apparent highness or lowness of a note, associated with frequency of vibration is called ____.
10. ______________ Tone color or tone quality is called ____.
11. ______________ A pleasant combination of pitches is called ____.

Name ______________________

15 Study Guide

Section 15.1: Properties of Sound

In your textbook, read about sound waves.
Write the term that correctly completes each statement.

Sound waves move in the same direction as the particles of the medium and are therefore **(1)** ____________________ waves. The waves are caused by variations in **(2)** ____________________ relating to the different **(3)** ____________________ of molecules. Therefore, sound cannot travel through a(n) **(4)** ____________________. The **(5)** ____________________ of a sound wave is the number of pressure oscillations per second. The **(6)** ____________________ is the distance between successive regions of high or low pressure. The speed of a sound wave in air depends on the **(7)** ____________________ of the air. At 20°C, sound moves through air at sea level at a speed of **(8)** ____________________ m/s. In general, the speed of sound is **(9)** ____________________ in liquids and solids than in gases. Reflected sound waves are **(10)** ____________________. The reflection of sound waves can be used to find the **(11)** ____________________ between a source and a reflecting surface. Sound waves can **(12)** ____________________, producing nodes, where little sound is heard. Sound waves can also be **(13)** ____________________; they spread outward after passing through a narrow opening. The equation that relates velocity, frequency, and wavelength is **(14)** ____________________.

In your textbook, read about loudness.
For each of the statements below, write true *or rewrite the italicized part to make the statement true.*

15. Sound level is measured on a *logarithmic* scale. ____________________

16. The ear detects *pressure* variations as sound. ____________________

17. Hearing loss can result from long exposure to *dissonant* sound levels. ____________________

18. Sound level is measured in *hertz.* ____________________

19. Loudness, as perceived by the human ear, is directly proportional to *length* of a pressure wave.

____

20. A *40-dB* increase is perceived as roughly a doubling of the loudness of a sound. ____________________

Name ______________________________

15 Study Guide

In your textbook, read about sound waves and loudness.

For each term on the left, write the corresponding symbol from those on the right. Some symbols may be used more than once.

____________	**21.** period	dB
____________	**22.** amplitude	Hz
____________	**23.** speed	m
____________	**24.** frequency	m/s
____________	**25.** loudness	s
____________	**26.** pitch	
____________	**27.** wavelength	

In your textbook, read about pitch and the Doppler effect.

State whether the process would make the apparent pitch of a sound higher, lower, or have no effect on it.

28. moving the listener away from the sound source ______________________________

29. decreasing the frequency ______________________________

30. increasing the number of decibels ______________________________

31. decreasing the wavelength, at constant wave speed ______________________________

32. increasing the wavelength, at constant wave speed ______________________________

33. moving the sound source toward the listener ______________________________

34. decreasing the wave amplitude ______________________________

35. increasing the wavelength, at constant frequency ______________________________

36. decreasing the period ______________________________

In your textbook, read about the Doppler shift.

The drawing below shows the crests of sound waves produced by a point source, S, that is moving. The two ears represent two listeners at different positions. Refer to the drawing to answer the following questions.

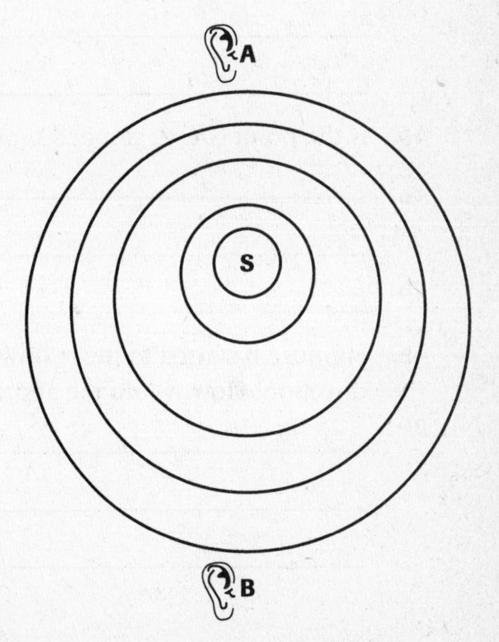

37. Which direction is the point source moving? Give a reason for your answer.

____

____

____

38. How do the apparent wavelengths at A and B compare?

____

____

Name ______________________

15 Study Guide

39. How does the speed of the sound waves compare at A and B?

40. How does the apparent frequency at point A compare to that at point B? Give a reason for your answer, in terms of the meaning of the word *frequency* and of the equation relating frequency to speed and wavelength.

41. How loud is the sound heard at A compared to that heard at B when A and B are equally far from the source? Give a reason for your answer.

42. Is the sound heard at A higher or lower pitch than that heard at B? Give a reason for your answer.

43. If the point source began to move faster in the same direction, would the apparent sound at A and B change? If yes, how would it change?

44. If the point source stopped moving, how would the apparent sound at A and B change, if at all?

45. Suppose B started to move downward while the point source continued to move in its original direction. How would the apparent sound at B change, if at all?

Name ______________________________

15 Study Guide

Section 15.2: The Physics of Music

In your textbook, read about pipe resonators and other sources of sound.
Circle the letter of the choice that best completes the statement or answers the question.

1. Sound is produced when there are ______.

a. increases in pressure
b. oscillations in pressure
c. increases in temperature
d. electromagnetic waves

2. The frequencies of vibrating air that will be set into resonance are determined by an air column's ______.

a. radius **b.** length **c.** mass **d.** width

3. Resonance occurs when ______.

a. any constructive interference occurs
b. any destructive interference occurs
c. a standing wave is created.
d. no nodes are formed

4. In which kind of resonator is the pressure of the reflected wave inverted?

a. closed-pipe only
b. open-pipe only
c. both open- and closed-pipe
d. neither open- nor closed-pipe

5. In a standing sound wave in a pipe, nodes are regions of ______.

a. maximum or minimum pressure and low displacement
b. maximum or minimum pressure and high displacement
c. mean atmospheric pressure and low displacement
d. mean atmospheric pressure and high displacement

6. In a standing sound wave in a pipe, two antinodes are separated by ______.

a. one-quarter wavelength
b. one wavelength
c. one-half wavelength
d. two wavelengths

In your textbook, read about resonance frequencies in pipes.
For each of the statements below, write true *or* false.

______________ **7.** A closed pipe can't have resonance unless it has antinodes at both ends.

______________ **8.** In a closed pipe, a column of length $\lambda/4$ will be in resonance with a tuning fork.

______________ **9.** An open pipe can't have resonance unless it has nodes at both ends.

______________ **10.** In an open pipe, a column of length $3\lambda/4$ will be in resonance with a tuning fork.

______________ **11.** For both open and closed pipes, resonance lengths are spaced at half-wavelength intervals.

Name ______________________

15 Study Guide

In your textbook, read about sound detectors.

Answer the following questions, using complete sentences.

12. Briefly describe energy conversion by a microphone.

____

____

____

13. Describe the pathway that sound waves follow as they enter the ear and eventually lead to the sensation of sound.

____

____

____

____

In your textbook, read about sound qualities and beat.

For each of the statements below, write true *or rewrite the italicized part to make the statement true.*

14. ______________ The *highest* frequency that will resonate in a musical instrument is called the fundamental.

15. ______________ In a *closed-pipe* instrument, such as a clarinet, the harmonics are odd-number multiples of the fundamental frequency.

16. ______________ A sound spectrum of an instrument is a graph of wave amplitude versus *speed.*

17. ______________ Sounds perceived as *consonant* in most Western cultures tend to have frequencies that are in small, whole-number ratios.

18. ______________ Interference between two nearly identical frequencies causes a *consonance.*

19. ______________ A beat is an oscillation of wave amplitude.

20. ______________ The frequency of the beat equals the magnitude of *difference* between the frequencies of the two waves.

In your textbook, read about musical intervals.

Write the frequency ratio for the pitches in each musical interval.

21. octave _____

22. fifth _____

23. major third _____

24. fourth _____

Date ______________ Period ______________ Name ______________________

16 Study Guide

Use with Chapter 16.

Light

Vocabulary Review

Write the term that correctly completes each statement. Use each term once.

complementary color	luminous	polarized	secondary pigment
dye	luminous flux	primary pigment	spectrum
illuminance	luminous intensity	ray model	translucent
illuminated	opaque	secondary color	transparent
light	pigment		

1. ______________________ The illumination of a surface is _____.
2. ______________________ The _____ of a given color forms white light when it combines with that color.
3. ______________________ The _____ of a point source is the luminous flux that falls on 1 m^2 of a sphere with a 1-m radius.
4. ______________________ A measure of the total rate at which light is emitted from a source is _____.
5. ______________________ Light that vibrates in a particular plane is _____.
6. ______________________ A color formed from two primary colors of light is a(n) _____.
7. ______________________ The range of frequencies of electromagnetic waves that stimulates the retina of the eye is _____.
8. ______________________ The _____ uses straight lines to describe how light behaves.
9. ______________________ A material that transmits light without distorting it is _____.
10. ______________________ The ordered arrangement of colors from red through violet is the _____.
11. ______________________ A colored material that is made up of particles larger than molecules is a(n) _____.
12. ______________________ A molecule that absorbs certain wavelengths and transmits others is a(n) _____.
13. ______________________ A(n) _____ absorbs only one primary color from white light.
14. ______________________ A(n) _____ absorbs two primary colors of light.
15. ______________________ A material that transmits light, but objects cannot be seen clearly through it is a(n) _____ material.
16. ______________________ A material that transmits no incident light is a(n) _____ material.
17. ______________________ A(n) _____ body does not emit light but simply reflects light produced by a source.
18. ______________________ A(n) _____ body emits light waves.

Name ____________________

16 Study Guide

Section 16.1: Light Fundamentals

In your textbook, read about characteristics of light.

For each of the statements below, write true *or rewrite the italicized part to make the statement true.*

__________ **1.** Light waves have wavelengths from about *400* nm to about 700 nm.

__________ **2.** The shortest wavelengths of light are seen as *red* light.

__________ **3.** Light travels in a *straight* line in a vacuum.

__________ **4.** The speed of light in a vacuum is approximately 3.00×10^{10} *m/s*.

In your textbook, read about characteristics of light.

Circle the letter of the choice that best completes the statement or answers the question.

5. Light behaves like _____.

a. particles only
c. both particles and waves
b. waves only
d. neither particles nor waves

6. Geometric optics studies light by means of _____.

a. ray diagrams
c. ripple tanks
b. the wave model
d. a particle-wave model

7. The fact that sharp shadows are cast by the sun is an indication that light _____.

a. is a wave.
c. is made up of colors
b. travels in straight lines
d. travels in a curved path

8. A straight line that represents the path of a very narrow beam of light is

a. normal
c. a ray
b. a complement
d. a wave

9. The orbital period of which of the following was studied in the seventeenth century to determine the speed of light?

a. Io
c. Jupiter
b. Saturn
d. Earth's moon

10. How long does light take to cross Earth's orbit?

a. 2 s
c. 16 min
b. 8 min
d. 22 min

11. Light of which of the following colors has the longest wavelength?

a. red
c. blue
b. yellow
d. violet

Name ______________________

16 Study Guide

In your textbook, read about measures of light.
Complete the table below.

Table 1

Quantity	Name of Unit	Symbol for Unit
illuminance		
luminous flux		
luminous intensity		

In your textbook, read about sources of light.
Write the term that correctly completes each statement.

A body that **(12)** ______________________ light waves is luminous, and a body that simply **(13)** ______________________ light waves given off by a source is illuminated. A(n) **(14)** ______________________ object emits light as a result of its high temperature. The official SI unit from which all light intensity units are calculated is the **(15)** ______________________. Luminous flux expresses the **(16)** ______________________ at which light is given off. Illuminance is the **(17)** ______________________ of a surface. If the distance from a surface to a point source of light is doubled, the illumination reaching the surface is only **(18)** ______________________ as great. The relationship between illuminance and luminous flux is given by the equation **(19)** ______________________, where the distance to the surface is represented by the symbol **(20)** ______________________. This equation is valid only if the light from the source is traveling **(21)** ______________________ to the surface. It is also valid only for sources small enough or far away enough to be **(22)** ______________________ sources.

In your textbook, read about light sources and quantities.
For each situation below, write the letter of the matching item.

_____ **23.** symbol for the speed of light
_____ **24.** area of the surface of a sphere
_____ **25.** symbol for luminous flux
_____ **26.** symbol for illuminance
_____ **27.** equivalent to the speed of light
_____ **28.** equivalent to 1 lx
_____ **29.** proportional to illumination
_____ **30.** equivalent to luminous intensity

a. $1\ \text{lm/m}^2$
b. $1/r^2$
c. $4\pi r^2$
d. c
e. E
f. λf
g. P
h. $P/4\pi$

Name ______________________________

16 Study Guide

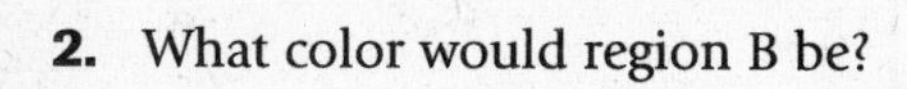

Section 16.2: Light and Matter

In your textbook, read about the additive color process.

The diagram below represents three overlapping circles of equally intense light of different pure colors. Assume that the circles are projected onto a white screen in an otherwise completely dark room.

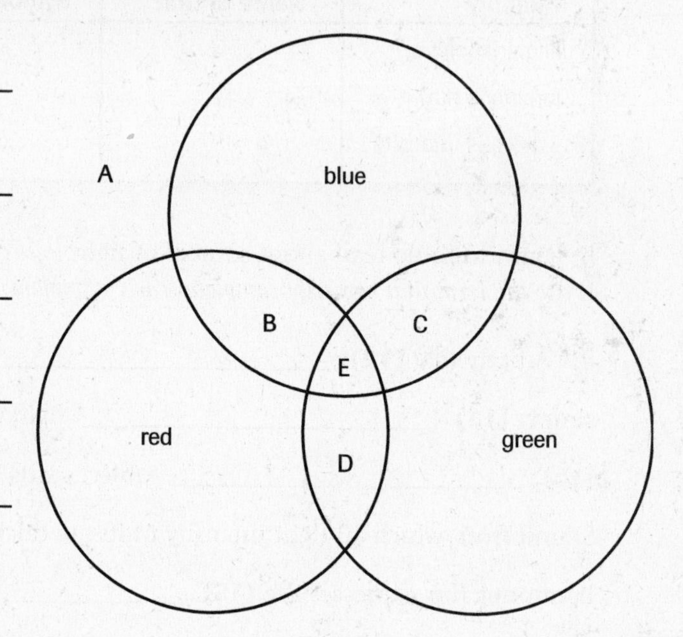

1. What color (if any) would region A be?

____

2. What color would region B be?

____

3. What color would region C be?

____

4. What color would region D be?

____

5. What color would region E be?

____

6. What color types of light—primary or secondary—are red, blue, and green light?

____

7. What color type of light would be in region B? Give a reason for your answer.

____

8. What color type of light would be in region C?

____

9. What color type of light would be in region D?

____

10. The color in region B is the complementary color to which color?

____

11. What happens if two complementary colors are projected together at the correct intensities onto a white screen?

____

12. The color in region C is complementary to which color?

____

13. The color in region D is complementary to which color?

____

Name ______________________________

16 Study Guide

In your textbook, read about pigments.

Answer the following questions, using complete sentences.

14. What does a primary pigment absorb from white light?

____

15. What does a secondary pigment absorb from white light?

____

16. Is the absorption of light to form pigment colors an additive or a subtractive process?

____

17. What color are the three primary pigments?

____

18. What color are the three secondary pigments?

____

19. What color(s) does yellow pigment absorb from light?

____

20. What color(s) does yellow pigment reflect?

____

21. What colors does red pigment absorb from light?

____

22. What is the complementary pigment to cyan pigment?

____

23. What color would result from the mixing of two complementary pigments?

____

In your textbook, read about thin films.

For each of the statements below, write true *or rewrite the italicized part to make the statement true.*

24. ____________________ The colors produced by an oil film on water result from thin-film *interference.*

25. ____________________ When a soap film is held vertically, it is *thicker* at the top than at the bottom.

26. ____________________ When a light wave traveling through the air reaches the top surface of a soap film, the wave that is reflected there is *inverted.*

27. ____________________ When a light wave that has traveled into a soap film reaches the back surface, the wave that is reflected there is *inverted.*

28. ____________________ When a soap film has a thickness of $\lambda/4$, the wave reflected from the back surface *destructively interferes with* the wave reflected from the top surface.

Name ______________________

16 Study Guide

29. ______________________ When a soap film has a thickness of $5\lambda/4$, the wave reflected from the back surface *reinforces* the wave reflected from the top surface.

30. ______________________ In a soap film, the $\lambda/4$ requirement is met at the same locations for *all colors* of light.

31. ______________________ The explanation of thin-film interference does not require a *wave* model of light.

In your textbook, read about polarization.

For each of the statements below, write true *or rewrite the italicized part to make the statement true.*

32. Light that has been polarized vibrates in *all planes.*

____

33. Polaroid material allows electromagnetic waves vibrating in one direction to pass through, while absorbing waves vibrating in the *perpendicular* direction.

____

34. Light can be polarized by *refraction.*

____

35. Photographers use polarizing filters to block *reflected* light.

____

Date ______________ Period ______________ Name ______________________

17 Study Guide

Use with Chapter 17.

Reflection and Refraction

Vocabulary Review

Write the term that correctly completes each statement. Use each term once.

angle of incidence	**dispersion**	**refraction**
angle of reflection	**index of refraction**	**regular reflection**
angle of refraction	**normal**	**Snell's law**
critical angle	**optically dense**	**total internal reflection**
diffuse reflection		

1. ______________ When light bounces off surfaces that are not very smooth, _____ occurs.

2. ______________ A line perpendicular to a surface is a(n) _____.

3. ______________ When light passes from one medium to a less optically dense medium at an angle so great that there is no refracted ray, _____ occurs.

4. ______________ The separation of light into its spectrum is _____.

5. ______________ When light bounces off surfaces that are extremely smooth, so that the light returns to the observer in parallel beams, _____ occurs.

6. ______________ The angle that an incoming beam makes with the normal is the _____.

7. ______________ The relationship of the angle of incidence to the angle of refraction is stated in _____.

8. ______________ If the angle of refraction is smaller than the angle of incidence, the medium in which the angle is smaller is more _____.

9. ______________ The angle that a refracted ray makes with the normal is the _____.

10. ______________ For light going from a vacuum into another medium, the constant n is the _____.

11. ______________ The angle made with the normal by a beam of light that has bounced off a surface is the _____.

12. ______________ The bending of light at the boundary between two media is _____.

13. ______________ The _____ describes a refracted ray that lies along the boundary of a substance.

Name ______________________________

17 Study Guide

Section 17.1: How Light Behaves at a Boundary

In your textbook, read about reflection and refraction.

For each of the statements below, write true *or rewrite the italicized part to make it true.*

1. ______________________ The angle of incidence is the angle between the incoming light beam and the *surface.*

2. ______________________ The velocity of light *does not* depend on the medium through which the light is traveling.

3. ______________________ The angle of incidence equals the angle of *reflection.*

4. ______________________ A mirror causes *diffuse reflection.*

5. ______________________ *Reflection* involves a bending of light.

6. ______________________ The angle of incidence *differs* from the angle of refraction.

Write the term that correctly completes each statement.

When a beam of light strikes a rough surface, it reflects at **(7)** ______________________ angles, producing **(8)** ______________________ reflection. When a beam of light strikes a very smooth surface, the reflected rays are **(9)** ______________________ to each other, producing **(10)** ______________________ reflection. If a light beam is not reflected, but is bent instead, the light is **(11)** ______________________. The angle of incidence is measured between the **(12)** ______________________ ray and the **(13)** ______________________. The angle of refraction is measured between the **(14)** ______________________ ray and the **(15)** ______________________. Refraction does not occur if the angle of incidence is **(16)** ______________________ degrees. When a light ray enters a medium that is more **(17)** ______________________ dense, its speed **(18)** ______________________, and the refracted ray bends **(19)** ______________________ the normal. If a light ray passes into a medium in which it travels faster, the refracted ray bends **(20)** ______________________ the normal.

Name ____________________

17 Study Guide

In your textbook, read about Snell's law and the effect of different media on light.

The diagram below represents a light ray passing through two media. Refer to the diagram to answer the following questions. Use complete sentences.

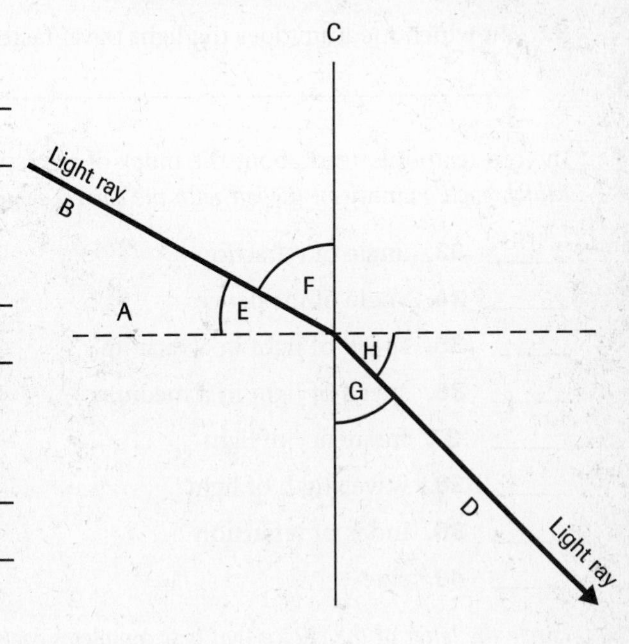

21. Which labeled line represents the incident ray? How can you tell?

____

____

22. Which line represents the refracted ray? How can you tell?

____

____

23. Which line represents the boundary between the two media?

____

____

24. Which line represents the normal? How can you tell?

____

25. Which labeled angle represents the angle of incidence? How can you tell?

____

26. Which labeled angle represents the angle of refraction? How can you tell?

____

27. Which is greater: the angle of incidence or the angle of refraction?

____

28. Which medium is more optically dense? How can you tell?

____

____

29. Which medium has the lower refractive index? How can you tell?

____

30. Write an equation for Snell's law.

____

Name ______________________________

17 Study Guide

31. Using Snell's law, complete the following equation, showing the ratio of the indices of refraction for medium 1 and medium 2: n_2/n_1 = _____.

____

32. In which medium does the light travel faster? How can you tell?

____

In your textbook, read about the index of refraction and speed of light.
Match each quantity on the left with the correct symbol.

_____	**33.** angle of refraction	c
_____	**34.** angle of incidence	θ_r
_____	**35.** speed of light in a vacuum	n
_____	**36.** speed of light in a medium	sin
_____	**37.** frequency of light	f
_____	**38.** wavelength of light	θ_i
_____	**39.** index of refraction	v
_____	**40.** sine	λ

Circle the letter of the choice that best completes each statement.

41. Compared to its speed in air, the speed of light in glass is _____.

a. the same **b.** faster. **c.** slower **d.** not measurable

42. Which of the following has the highest index of refraction? _____

a. a vacuum **b.** air **c.** diamond **d.** crown glass

43. For most practical purposes, the index of refraction of air can be considered _____.

a. 0 **c.** infinite
b. 1.00 **d.** equal to that of the other medium involved

44. The wavelength of light in a given material equals _____.

a. the speed of light in the material times frequency
b. frequency times the speed of light in the material
c. the speed of light in the material divided by frequency
d. frequency divided by the speed of light in the material

45. Which expression equals the index of refraction of a given medium?

a. c/v_{medium} **b.** v_{medium}/c **c.** $c - v_{medium}$ **d.** $v_{medium} - c$

Name ______________________________

17 Study Guide

Section 17.2: Applications of Reflected and Refracted Light

In your textbook, read about total internal reflection.
For each of the statements below, write true *or* false.

_________ **1.** The critical angle is the incident angle that causes the refracted ray to lie along the boundary.

_________ **2.** The sine of the critical angle for a substance from which light is passing into air is equal to $1.00/n_i$.

_________ **3.** Total internal reflection cannot be achieved for light that is passing from water to air.

_________ **4.** All substances have the same critical angle.

_________ **5.** Plants can use total internal reflection for light transport.

_________ **6.** A single bundle of optical fibers can carry thousands of telephone conversations at once.

_________ **7.** Total internal reflection occurs whenever light passes from one medium to a less optically dense medium.

_________ **8.** When incident light is at the critical angle, θ_r equals $0°$.

_________ **9.** The core of an optical fiber is metal.

_________ **10.** Each time light moving through an optical fiber strikes the surface, the angle of incidence is larger than the critical angle.

_________ **11.** Cladding covers the outside of an optical fiber.

_________ **12.** Optical fibers are flexible.

In your textbook, read about the effects of refraction.
Circle the letter of the choice that best completes the statement or answers the question.

13. As temperature increases, the index of refraction of air _____.

a. increases
b. decreases
c. remains constant
d. becomes negligible

14. The mirage effect is caused by _____.

a. total internal reflection
b. polarization
c. temperature differences in index of refraction
d. differences in the speeds of different colors of light

15. Which of the following is correct regarding sunlight? _____

a. It reaches us only after the sun has actually risen.
b. It continues to reach us after the sun has actually set.
c. It stops reaching us before the sun has actually set.
d. It corresponds exactly to sunrise and sunset.

Name ______________________

17 Study Guide

16. When all the colors of light are combined, _____.
    - **a.** black results
    - **b.** the frequencies add together, producing light of higher frequency
    - **c.** the colors completely cancel out because of destructive interference
    - **d.** white light is produced

17. Which color of light is bent most as it goes through a prism?

    a. red **b.** yellow **c.** green **d.** violet

18. The color of light that has the lowest index of refraction is _____.

    a. red **b.** yellow **c.** green **d.** violet

19. Diamond has _____.
    - **a.** a low refractive index and little color variation in the index
    - **b.** a low refractive index and lots of color variation in the index
    - **c.** a high refractive index and little color variation in the index
    - **d.** a high refractive index and lots of color variation in the index

20. Which description fits the spectrum of light from an incandescent lamp?
    - **a.** a continuous band only
    - **b.** bright lines only
    - **c.** both a continuous band and bright lines
    - **d.** neither a continuous band nor bright lines

21. Which description fits the spectrum of light from a fluorescent lamp?
    - **a.** a continuous band only
    - **b.** bright lines only
    - **c.** both a continuous band and bright lines
    - **d.** neither a continuous band nor bright lines

22. Production of a rainbow by water droplets depends on _____.
    - **a.** differences of angles of incidence for lights of different colors
    - **b.** differences of angles of refraction for lights of different colors
    - **c.** differences of speed for lights of different colors
    - **d.** equality of index of refraction for lights of all colors

23. A puddle mirage seen by a motorist is actually light from _____.
    - **a.** the ground underneath the puddle
    - **b.** the ground surrounding the puddle
    - **c.** the sky.
    - **d.** passing vehicles.

Date ______________ Period ______________ Name ______________________________

18 Study Guide

Use with Chapter 18.

Mirrors and Lenses

Vocabulary Review

Write the term that correctly completes each statement. Use each term once.

achromatic lens	convex lens	focal point	principal axis
chromatic aberration	convex mirror	lens/mirror equation	real image
concave lens	erect image	magnification	spherical aberration
concave mirror	focal length	object	virtual image

1. ______________________ An image at which light rays actually converge is a(n) _____.

2. ______________________ A(n) _____ reflects light from its inwardly curving surface.

3. ______________________ The straight line perpendicular to the surface of a mirror at its center is the _____.

4. ______________________ The mathematical relationship between focal length, distance of object, and distance of image is expressed by the _____.

5. ______________________ A transparent refracting device that is thinner in the middle than at the edges is a(n) _____.

6. ______________________ A transparent refracting device that is thicker in the middle than at the edges is a(n) _____.

7. ______________________ The distance between the focal point and the mirror or lens is the _____.

8. ______________________ Any source of diverging light rays is a(n) _____.

9. ______________________ An image at which light rays do not actually converge is a(n) _____.

10. ______________________ The place at which light rays parallel to the principal axis of a concave mirror converge is the _____.

11. ______________________ An undesirable effect in which an object viewed through a lens appears to be ringed with color is _____.

12. ______________________ An image that is not inverted is a(n) _____.

13. ______________________ An undesirable effect in which the parallel rays reflected in a concave mirror fail to meet at a point is _____.

14. ______________________ A spherical mirror that reflects light from its outwardly curving surface is a(n) _____.

15. ______________________ A lens constructed so as to avoid undesirable color effects is a(n) _____.

16. ______________________ The ratio of the size of an image to the size of the object that produces it is the _____.

Name ______________________________

18 Study Guide

Section 18.1: Mirrors

In your textbook, read about concave mirrors.

For each of the statements below, write true *or rewrite the italicized part to make the statement true.*

1. ______________________ Rays *perpendicular* to the principal axis of a concave mirror converge at or near the focal point.

2. ______________________ The focal length of a concave mirror is *half* the radius of curvature.

3. ______________________ If the object is *farther out* than the center of curvature of a concave mirror, its image appears between the focus and the center of curvature.

4. ______________________ Concave mirrors can produce *only virtual* images.

5. ______________________ Concave mirrors *cannot* act as magnifiers.

In your textbook, read about real images formed by concave mirrors.

Circle the letter of the choice that best answers each question.

6. Which of the following correctly states the lens/mirror equation?

a. $f = d_i + d_o$ **c.** $f = 1/d_i + 1/d_o$

b. $1/f = d_i + d_o$ **d.** $1/f = 1/d_i + 1/d_o$

7. Which of the following is a correct relationship?

a. $m = h_i + h_o$ **c.** $m = h_i/h_o$

b. $m = h_i - h_o$ **d.** $m = h_o/h_i$

8. Which of the following indicates that an image produced by a concave mirror is upright?

a. a positive value for h_i **c.** a positive value for d_i

b. a negative value for h_i **d.** a negative value for d_i

9. To which of the following is m equal?

a. d_o/d_i **c.** d_i/d_o

b. $-d_o/d_i$ **d.** $-d_i/d_o$

In your textbook, read about virtual images and image defects in concave mirrors.

Circle the letter of the choice that best answers each question.

10. Which of the following indicates that an image produced by a concave mirror is virtual?

a. a positive value for h_i **c.** a positive value for d_i

b. a negative value for h_i **d.** a negative value for d_i

11. If an object is placed at the focal point of a concave mirror, where will the image be?

a. also at the focal point **c.** at infinity

b. at the center of curvature **d.** at the surface of the mirror

Name ________________________

18 Study Guide

12. Which of the following posed a problem for the Hubble Space Telescope?

a. improperly ground lenses

b. chromatic aberration

c. spherical aberration

d. cracked spherical mirrors

13. Why don't parabolic mirrors have trouble with spherical aberration?

a. All parallel rays are reflected to a single spot.

b. All parallel rays focus on infinity.

c. They use a secondary mirror for correction.

d. They have a virtual focus point.

In your textbook, read about real images formed by concave mirrors.

The diagram below shows a concave mirror and an object. Refer to the diagram to answer the following questions.

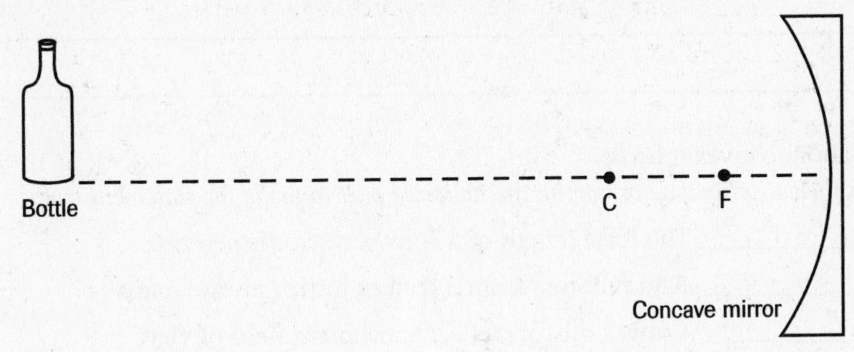

14. What does the letter *F* stand for?

____

15. What does the dashed line represent?

____

16. Under what condition will rays be reflected through point F?

____

17. Relative to points C and F, roughly where will the image be?

____

18. Will the value of d_i be positive or negative? Why?

____

19. Will the image be real or virtual? Why?

____

20. Will the image be reduced or enlarged?

____

21. Will the value of h_i be greater or less than that of h_o?

____

22. Will the image be erect or inverted?

____

23. Will the value of h_i be positive or negative? Why?

____

24. If the object was moved closer to C but still remained to the left of it, what would happen to the image?

____

25. What would happen to the image if the object was moved to C?

____

____

In your textbook, read about convex mirrors.

For each of the statements below, write true *or rewrite the italicized part to make the statement true.*

26. ______________________ The focal length of a convex mirror is *negative.*

27. ______________________ Rays reflected from a convex mirror always *converge.*

28. ______________________ Convex mirrors reflect an *enlarged* field of view.

29. ______________________ The images produced by convex mirrors are *real* images.

30. ______________________ When the magnification is negative, an image will be *erect.*

31. ______________________ Compared to the size of the corresponding objects, the images produced by convex mirrors are always *the same size.*

Section 18.2: Lenses

In your textbook, read about real images formed by convex lenses.

Circle the letter of the choice that best completes the statement or answers the question.

1. Convex lenses typically have _____.
   - **a.** one focal point
   - **b.** two focal points
   - **c.** a virtual focal point
   - **d.** no focal point
2. For a convex lens, if an object is between f and $2f$, the image will be _____.
   - **a.** smaller than the object
   - **b.** larger than the object
   - **c.** the same size as the object
   - **d.** at infinity
3. For a lens, which of the following relationships is correct?
   - **a.** $m = h_o/h_i$
   - **b.** $m = d_i/d_o$
   - **c.** $m = -h_o/h_i$
   - **d.** $m = -d_i/d_o$

Name ______________________

18 Study Guide

4. The relationship $1/f = 1/d_i + 1/d_o$ is ______.
   - **a.** valid for any lens
   - **b.** not valid for lenses
   - **c.** valid for convex lenses only
   - **d.** valid for concave lenses only
5. What kinds of images can convex lenses produce?
   - **a.** real only
   - **b.** virtual only
   - **c.** both real and virtual
   - **d.** neither real nor virtual
6. What is the principal advantage of using a large convex lens rather than a small one?
   - **a.** elimination of spherical aberration
   - **b.** elimination of chromatic aberration
   - **c.** increasing the size of the focal point
   - **d.** collection of more light rays
7. If an object is placed at the focal point of a convex lens, the refracted rays will ______.
   - **a.** emerge in a parallel beam
   - **b.** converge at the other focal point
   - **c.** converge at the lens surface
   - **d.** diverge

In your textbook, read about virtual images formed by convex lenses.
Answer the following questions, using complete sentences.

8. What type of lens can be used as a magnifying glass?

   ____

9. When a lens is used as a magnifying glass, where is the object placed?

   ____

10. When a lens is used as a magnifying glass, what sign(s) do d_i and h_i have?

    ____

In your textbook, read about concave lenses and lens defects.
For each of the statements below, write true *or rewrite the italicized part to make the statement true.*

11. ______________ The images produced by concave lenses are always *inverted.*
12. ______________ A concave lens is *thinner* in the center than at the edges.
13. ______________ Concave lenses refract light rays so that the rays *converge.*
14. ______________ The images produced by concave lenses are *virtual and enlarged.*
15. ______________ Lenses suffer from spherical aberration because the rays that pass through *do not* all pass through the focus.
16. ______________ The edges of a lens act like a *prism,* scattering light in a ring of color.
17. ______________ You can reduce *spherical aberration* by joining a concave lens with a convex lens.

Name ______________________________

18 Study Guide

In your textbook, read about microscopes and telescopes.

Circle the letter of the choice that best completes the statement or answers the question.

18. The objective lens of a microscope is used to produce an image located _____.

a. at the ocular lens

b. between the ocular lens and its focal point

c. at the focal point of the ocular lens

d. beyond the ocular lens and its focal point

19. What kind of image is a telescope designed to produce for the viewer?

a. real and inverted

b. real and erect

c. virtual and inverted

d. virtual and erect

20. An astronomical refracting telescope uses _____.

a. two concave lenses

b. a combination of concave and convex lenses

c. two convex lenses

d. a variety of lenses and mirrors

Date ______________ Period ______________ Name ______________________

19 Study Guide

Use with Chapter 19.

Diffraction and Interference of Light

Vocabulary Review

Write the term that correctly completes each statement. Use each term once.

coherent waves	**diffraction grating**	**monochromatic light**
constructive interference	**first-order line**	**Rayleigh criterion**
corpuscular model	**grating spectrometer**	**resolving power**
destructive interference	**interference fringes**	**wavelet**
diffraction		

1. ______________________ A(n) _____ has parallel slits that diffract light and form an interference pattern.

2. ______________________ Light of only one wavelength is _____.

3. ______________________ The slight spreading of light around barriers is _____.

4. ______________________ Waves that are in phase are _____.

5. ______________________ The patterns of bright and dark bands produced when light passes through a double slit are _____.

6. ______________________ The _____ indicates whether two closely spaced star images will be resolved.

7. ______________________ The ability of an instrument, such as a telescope, to reveal the separateness of closely spaced images is its _____.

8. ______________________ When crests of waves overlap, _____ occurs.

9. ______________________ A bright band directly on either side of the central band is a(n) _____.

10. ______________________ A repeating disturbance produced when the crest of a wave is replaced by a wave source is a(n) _____.

11. ______________________ When crests of one wave overlap troughs of another wave, _____ occurs.

12. ______________________ An instrument used to measure light wavelengths produced by a diffraction grating is a(n) _____.

13. ______________________ The theory that light is made up of tiny particles is called the _____.

Name ______________________

19 Study Guide

Section 19.1: When Light Waves Interfere

In your textbook, read about diffraction and the properties of light.

Circle the letter of the choice that best completes the statement or answers the question.

1. One sign that the corpuscular model of light isn't entirely correct is that light _____.
   - a. travels in absolutely straight lines
   - b. produces fuzzy shadow edges
   - c. produces sharp shadow edges
   - d. adds energy to matter
2. The crest of any wave can be replaced by a series of _____.
   - a. equally spaced wave sources that are in step with each other
   - b. equally spaced wave sources that interfere with each other
   - c. unequally spaced wave sources that are in step with each other
   - d. unequally spaced wave sources that interfere with each other
3. Interference fringes result from _____.
   - a. constructive and destructive interference
   - b. particle interactions
   - c. wavelet creation
   - d. refraction
4. A monochromatic light source emits light of _____.
   - a. all frequencies
   - b. one wavelength
   - c. different wavelengths that interfere constructively
   - d. one refractive index
5. How many slits were present in the barrier Young used to observe interference fringes?
   - a. none
   - b. one
   - c. two
   - d. three
6. Wave crests that reach the same points at the same times are said to be _____.
   - a. out of phase
   - b. in phase
   - c. noncoherent
   - d. diffractive
7. Where two wave crests overlap, _____.
   - a. a bright band is created
   - b. a dark band is created
   - c. diffraction occurs
   - d. a colored band is created
8. Where a wave crest and a wave trough overlap, _____.
   - a. a bright band is created
   - b. a dark band is created
   - c. diffraction occurs
   - d. a colored band is created
9. In a two-slit experiment that uses monochromatic light, what appears at the center of the screen?
   - a. bands of two different colors
   - b. a dark band
   - c. a complete spectrum
   - d. a bright band

Name ______________________

19 Study Guide

10. In a two-slit experiment that uses white light, what appears at the center of the screen?

a. bands of two different colors

b. a dark band

c. a complete spectrum

d. a bright band

11. In a two-slit experiment that uses white light, what types of bands appear away from the center of the screen?

a. bright bands only

b. dark bands only

c. colored spectra

d. monochromatic bands

In your textbook, read about measuring wavelength with two-slit diffraction.
For each term on the left, write the matching item.

_____	**12.** width of a single slit	d
_____	**13.** wavelength	L
_____	**14.** separation of two slits	λ
_____	**15.** distance between the screen and slits	P
_____	**16.** location of the central bright band	P_0
_____	**17.** location of the first-order line	w
_____	**18.** distance between the central bright line and first-order line	x

The diagram below shows a two-slit experimental setup. Refer to the diagram to answer the following questions. Use complete sentences.

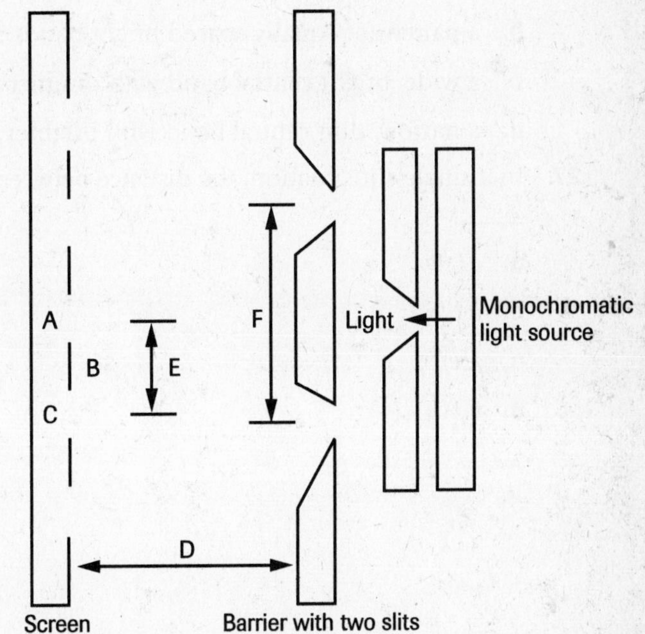

19. Describe the band that appears on the screen at point A. What kind of interference is occurring there?

____

____

20. What kind of band appears at point B? What kind of interference is occurring there?

____

____

21. Describe the band that appears on the screen at point C. What kind of interference is occurring there?

____

____

19 Study Guide

22. What does the length E represent?

____

23. Explain how the wavelength of the light can be calculated from the two-slit setup.

____

____

24. Write an equation calculating wavelength, first using the labels from the diagram, and then using the standard symbols.

____

In your textbook, read about single-slit diffraction.

Circle the letter of the choice that best completes the statement or answers the question.

25. Which of the following is true of the diffraction of light compared to that of sound?

a. Light is diffracted more because it has smaller wavelengths.

b. Light is diffracted more because it has larger wavelengths.

c. Sound is diffracted more because it has smaller wavelengths.

d. Sound is diffracted more because it has larger wavelengths.

26. If light is passed through a single slit, _____.

a. it is not diffracted

b. a pattern of equally spaced bright bands results

c. a wide, bright central band with dimmer bands on either side is produced

d. a narrow, dim central band with brighter bands on either side is produced

27. In a single-slit situation, the distance between the central band and the first dark band is equal to _____.

a. $\lambda L/w$

b. wL/λ

c. $\lambda w/L$

d. $w/\lambda L$

Name ______________________________

19 Study Guide

Section 19.2: Applications of Diffraction

In your textbook, read about diffraction gratings.

For each of the statements below, write true *or rewrite the italicized part to make the statement true.*

1. ______________________ When white light falls on a diffraction grating, *white bands* are produced.

2. ______________________ Diffraction gratings are made by using a diamond to scratch fine lines *on glass.*

3. ______________________ The spacing between the lines on diffraction gratings is typically about *1 mm.*

4. ______________________ Wavelength *cannot* be measured as precisely with diffraction gratings as with double slits.

Write the term that correctly completes each statement. Use each term once.

θ	**colors**	**interference**
λ/d	**diffract**	**narrower**
angle	**distance**	**wavelengths**
broader	**grating spectrometer**	

A diffraction grating has many parallel slits that **(5)** ______________________ light and form a(n) **(6)** ______________________ pattern. This pattern has bright bands that are **(7)** ______________________ than those produced by a double slit, and the dark regions are **(8)** ______________________. The pattern makes it easier to distinguish individual **(9)** ______________________ and to measure **(10)** ______________________ more precisely. The equation used in the calculation states that x/L is equal to **(11)** ______________________, where the variable in the denominator is the **(12)** ______________________ between the grating's lines. Instead of measuring the distance from the central band to the first bright band, most laboratory instruments measure the **(13)** ______________________ between them. The symbol for that quantity is **(14)** ______________________. The instrument used for the measurement is called a **(15)** ______________________.

Name ______________________

19 Study Guide

In your textbook, read about determining wavelength with diffraction gratings.

Circle the letter of the choice that best completes the statement or answers the question.

16. For a diffraction grating, the wavelength equals _____.

a. $\sin d$ **b.** $\sin d\theta$ **c.** $d/\sin \theta$ **d.** $d \sin \theta$

17. When monochromatic light is used with a diffraction grating, what is produced?

a. no diffraction pattern

b. no central band, but bright bands to the left and right

c. a bright central band with bright bands to the left and right

d. a dark central band with white bands to the left and right

18. How do the screen locations of the red lines produced by red light shone on a diffraction grating compare with the locations of the red lines produced by white light?

a. Those produced by the red light are above.

b. Those produced by the red light are to the left.

c. Those produced by the white light are closer together.

d. They are in the same places.

19. The sine of the angle between bright bands is approximately equal to _____.

a. d/L **b.** L/x **c.** x/L **d.** d/x

20. Which quantity is read directly from the calibrated base of a grating spectrometer?

a. θ **b.** λ **c.** f **d.** c

In your textbook, read about telescope and microscope resolution.

For each of the statements below, write true *or rewrite the italicized part to make the statement true.*

21. ______________________ Decreasing the width of a lens used in a telescope causes the diffraction pattern to become *narrower.*

22. ______________________ Two stars will be just resolved if the central bright band of one star falls on the *first dark band* of the second star.

23. ______________________ The diffraction pattern formed by blue light in a microscope is *wider* than that formed by red light.

24. ______________________ Using a larger lens is practical in reducing effects of diffraction on the resolving power of *both microscopes and telescopes.*

25. ______________________ *White* light would be more useful than green light in reducing diffraction patterns formed in a microscope.

26. ______________________ Telescope diffraction causes light from a star to appear *brighter.*

Date ______________ Period ______________ Name ______________________________

20 Study Guide

Use with Chapter 20.

Static Electricity

Vocabulary Review

Write the term that correctly completes each statement. Use each term once.

charged	coulomb	elementary charge
charging by conduction	Coulomb's law	insulators
charging by induction	electroscope	neutral
conductors	electrostatics	plasma

1. ______________________ Materials through which charges will not move easily are electrical _____.

2. ______________________ The study of electrical charges that can be collected and held in one place is _____.

3. ______________________ A(n) _____ is a device used to determine charge.

4. ______________________ The magnitude of the charge of an electron is the _____.

5. ______________________ Separating the charges in an object without touching the object is _____.

6. ______________________ Materials such as metals are electrical _____. They allow charges to move about easily.

7. ______________________ The positive charge in _____ objects exactly balances the negative charge.

8. ______________________ Giving a neutral body a charge by touching it with a charged body is _____.

9. ______________________ The magnitude of the force between charge q_A and charge q_B, separated by a distance d, is proportional to the magnitude of the charges and inversely proportional to the square of the distance; this is a statement of _____.

10. ______________________ The _____ is the SI standard unit of charge.

11. ______________________ A(n) _____ is a gaslike state of negatively charged electrons and positively or negatively charged ions.

12. ______________________ An object that exhibits electrical interaction after rubbing is said to be _____.

Name ______________________________

20 Study Guide

Section 20.1: Electrical Charge

In your textbook, read about charged objects.

Circle the letter of the choice that best completes each statement.

1. Electricity caused by rubbing is _____.

a. static electricity
b. lightning
c. current electricity
d. charge

2. An object that exhibits electrical interaction after rubbing is said to be _____.

a. positive
b. electrical
c. charged
d. negative

3. Two objects with the same type of charge _____.

a. have no affect on each other
b. repel each other
c. attract each other
d. have to be positive

4. Two objects with opposite charges _____.

a. have no affect on each other
b. repel each other
c. attract each other
d. have to be negative

5. Two types of charges are _____.

a. yellow and green
b. top and bottom
c. attractive and repulsive
d. positive and negative

In your textbook, read about charged objects.

Complete the chart below by marking the appropriate column for the charge on each material after it is rubbed.

Table 1

Material	Positively charged	Negatively charged
6. plastic		
7. wool		
8. hard rubber		
9. fur		
10. glass		

In your textbook, read about the microscopic reasons for charge.

Answer the following questions.

11. What are the negative and positive parts of an atom?

____

Name ______________________

20 Study Guide

12. How can an atom become charged?

____

13. What happens when two neutral objects, such as rubber and fur, are rubbed together?

____

In your textbook, read about conductors and insulators.

Decide whether the examples below are insulators or conductors. Mark the correct column.

Table 2

Example	Insulator	Conductor
14. a material through which a charge will not move easily		
15. glass		
16. air as a plasma		
17. aluminum		
18. an object, held at the midpoint and rubbed only one end, becomes charged only at the rubbed end		
19. copper		
20. dry wood		
21. a material through which charges move about easily		
22. graphite		
23. charges removed from one area are not replaced by charges from another area		
24. most plastics		
25. dry air		
26. charges applied to one area spread quickly over the entire object		

Section 20.2: Electrical Force

In your textbook, read about forces on charged bodies and lightning.

For each of the statements below, write true *or rewrite the italicized part to make the statement true.*

1. There are *two* kinds of electrical charges, positive and negative.

____

2. Charges *cannot* exert force on other charges over a distance.

____

Name ______________________

20 Study Guide

3. The force between two charged objects is *weaker* when the objects are closer together.

____

4. *Opposite* charges repel.

____

5. If an electroscope is given a *positive* charge, the leaves will spread.

____

6. Neutral objects can attract charged objects because of separation of charge in the *charged* object.

____

7. Lightning bolts *discharge* clouds.

____

In your textbook, read about electroscopes.
Refer to the drawings to answer the following questions.

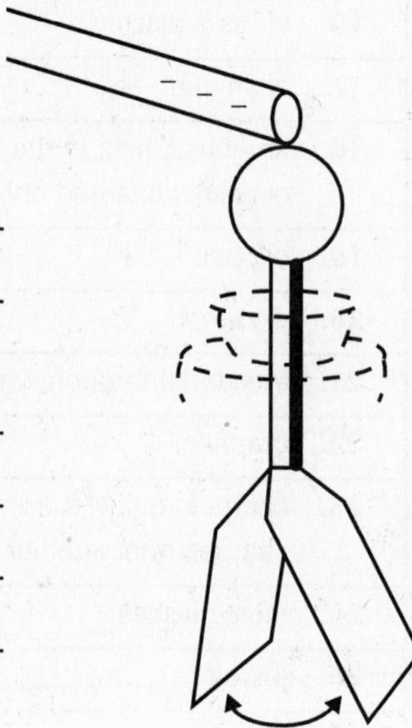

8. What is the net charge on the electroscope?

____

____

9. By what method is the electroscope being charged?

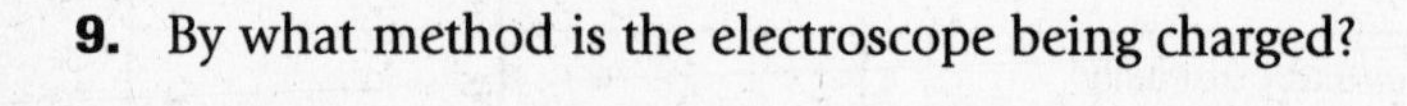

____

____

____

____

10. The electroscope has a net negative charge. What will happen if the electroscope is touched with an object that has a negative charge?

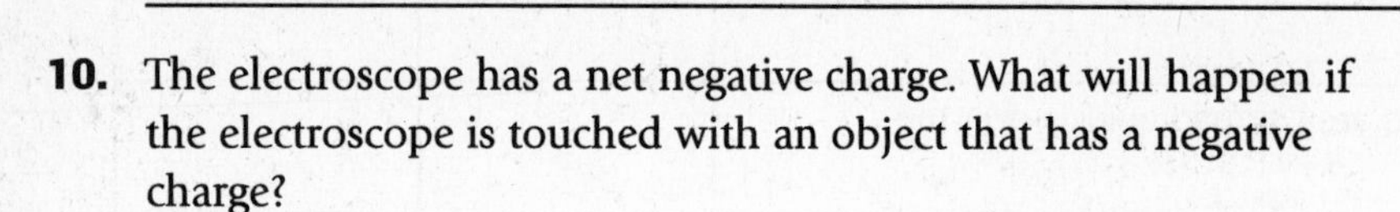

____

____

____

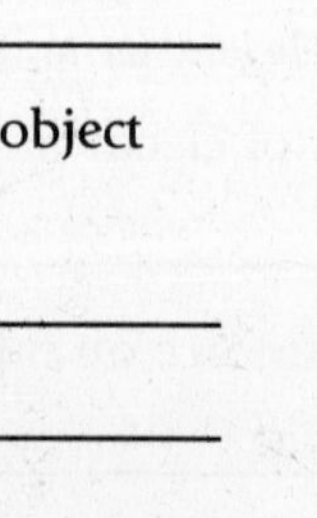

11. What will happen if the electroscope is touched with an object that has a positive charge?

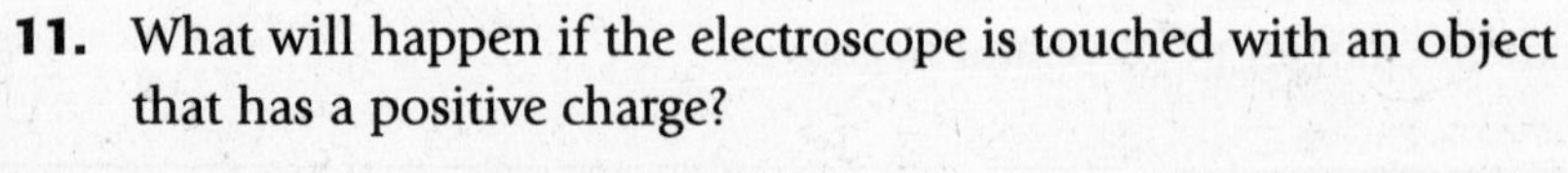

____

____

____

Name ______________________

20 Study Guide

12. What is the charge on the metal sphere A?

____

13. By what method are these metal spheres being charged?

____

In your textbook, read about Coulomb's law.

Circle the letter of the choice that best completes each statement.

According to Coulomb's law, the magnitude of the force on a charge q_A caused by charge q_B a distance d away can be written $F = K(q_A q_B/d^2)$.

14. The force, *F*, _____ with the square of the distance between the centers of two charged objects.

a. varies directly

b. varies inversely

c. varies negatively

d. doesn't vary

15. The force, *F*, _____ with the charge of two charged objects.

a. varies directly

b. varies inversely

c. varies negatively

d. doesn't vary

16. When the charges are measured in coulombs, the distance is measured in meters, and the force is measured in newtons, the constant, *K*, is _____

a. 1

b. 1.60×10^{-19} C

c. 9.0×10^{9} N·m^2/C^2

d. unknown

Name ______________________

20 Study Guide

17. Coulomb's law can be used to determine _____ of an electrical force.

a. the direction

b. the magnitude

c. the charge

d. both the magnitude and the direction

In your textbook, read about the application of electrical forces on neutral bodies.
Answer the following questions.

18. According to Newton's third law of motion, how is a neutral object affected by a charged object?

____

____

____

____

____

19. Give two examples of applications of electrical forces on neutral particles.

____

____

____

____

____

Date ______________ Period ______________ Name ______________________________

21 Study Guide

Use with Chapter 21.

Electric Fields

Vocabulary Review

Circle the letter of the choice that best completes each statement.

1. The unit used to measure the electric potential difference is the _____.

a. volt **c.** coulomb
b. voltmeter **d.** joule

2. The _____ of a charged object exerts force on other charged objects.

a. charge **c.** electric field
b. capacitance **d.** voltage

3. The ratio of charge to potential difference of an object is its _____.

a. capacitance **c.** voltage
b. capacity **d.** electric potential difference

4. The instrument that measures potential difference is the _____.

a. voltmeter **c.** potential difference meter
b. ammeter **d.** capacitor

5. The direction and strength of an electric field are depicted by _____ .

a. tangents **c.** spokes
b. electric field lines **d.** magnetic field lines

6. The change in potential energy per unit charge is _____.

a. volt **c.** electric potential difference
b. work **d.** capacitance

7. Touching an object to Earth to eliminate excess charge is _____.

a. connecting **c.** recharging
b. equalizing **d.** grounding

8. A device that stores charge by having a specific capacitance is a(n) _____.

a. charge holder **c.** capacitor
b. battery **d.** electric field

9. Capacitance is measured in _____.

a. coulombs **c.** ohms
b. volts **d.** farads

10. Electric charge may be visible as a(n) _____ on a power line.

a. corona **c.** electric vector
b. electric field line **d.** plasma

Name ______________________________

21 Study Guide

Section 21.1: Creating and Measuring Electric Fields

In your textbook, read about electric fields.
Answer the following questions, using complete sentences.

11. What produces an electric field?

____

12. Why must you use a test charge to observe an electric field?

____

13. How does Coulomb's law relate to test charges?

____

____

14. If arrows represent electric field vectors in a picture of an electric field, how are the magnitude and direction of the field shown?

____

____

15. How do you find the electric field from two charges?

____

____

16. Why should an electric field be measured only by a small test charge?

____

____

____

In your textbook, read about models for electric fields.
Answer the following questions.

17. Explain what a Van de Graaff machine is and one way it can be used to show field lines.

____

____

____

18. Do field lines and electric fields really exist? How are field lines and electric fields useful?

____

____

____

Name ______________________________

21 Study Guide

In your textbook, read about electric field vectors.
Refer to the illustration to answer the following questions.

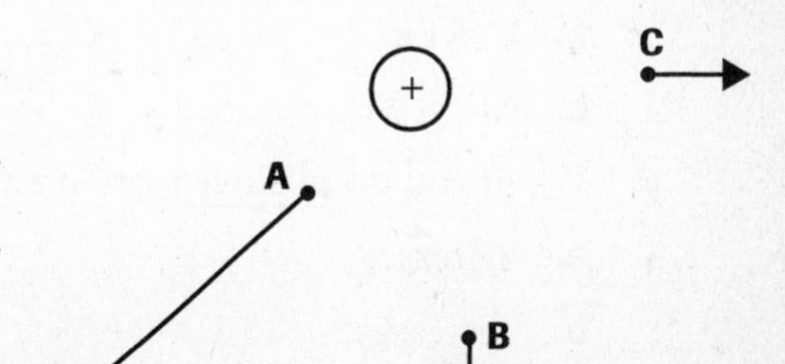

19. At what point is the magnitude of the electric field the greatest? Explain how you can tell from the drawing that this is true.

____

____

____

20. Points B and C are the same distance from the field charge. What could cause the force measured on a test charge at point B to be twice as large as the force measured on a test charge at point C?

____

____

____

21. If an arrow representing the electric field were drawn at point D, how long would it be relative to the other arrows in the drawing? Why?

____

____

In your textbook, read about electric field lines.
Answer the following questions.

22. Draw two electric charges, one positive and one negative. Assume they are so far apart that there is no interaction between the charges.

23. Draw electric field lines around each charge. Show that the field around the negative charge is twice as strong as the field around the positive charge.

Name ______________________

21 Study Guide

Section 21.2: Applications of Electric Fields

In your textbook, read about electric potential.

Circle the letter of the choice that best completes the statement or answers the question.

1. Which equation defines the electric potential difference?

a. $E = F/q'$
b. $\Delta V = \Delta V_e/q'$
c. $V = J/C$
d. $q = mg/E$

2. When you do positive work on a two-charge system, the electric potential energy _____.

a. increases
b. decreases
c. does not change
d. always disappears

3. Only _____ electric potential can be measured.

a. points of
b. absolute values of
c. differences in
d. attractions between

4. The electric potential _____ when a positive charge is moved toward a negative charge.

a. increases
b. stays the same
c. decreases
d. becomes positive

5. A positive test charge is located at point A. If the test charge is moved to some point B and then back to point A, what is the change in the electric potential?

a. The electric potential increases.
b. The electric potential decreases.
c. The electric potential becomes zero.
d. The electric potential does not change.

In your textbook, read about electric potential and uniform fields.

For each of the statements below, write true *or* rewrite *the italicized part to make the statement true.*

6. Two large, flat conducting plates parallel to each other can create a constant electric force and field. *Both are charged negatively.*

____

7. The direction of an electric field between two parallel conducting plates is *from the positive plate to the negative plate.*

____

8. The *electrical* difference between two points a distance d apart in a uniform field is represented by the equation $\Delta V = Ed$.

____

9. The potential increases in the direction *opposite* the electric field direction.

____

10. The potential is *lower* near the positively charged plate.

____

Name ____________________

21 Study Guide

In your textbook, read about Millikan's oil-drop experiment.

Write the term that correctly completes each statement. Use each term once.

charged	**gravitational field**	**multiple**	**suspended**
electric field	**ionized**	**potential difference**	**uniform**
electron	**magnitude**		

Millikan used a(n) **(11)** ____________________ electric field between two parallel plates to measure the charge on an electron. **(12)** ____________________ oil drops between the plates fell from the air unless the **(13)** ____________________ between the two plates was adjusted. When the top plate was positive enough, an oil drop was **(14)** ____________________ between the plates. At this adjustment, the force of Earth's **(15)** ____________________ and the force of the **(16)** ____________________ were the same magnitude. From the **(17)** ____________________ of the electric field, the charge on the drop was calculated. When Millikan **(18)** ____________________ the air to add or remove electrons from the drops, the change in the charge of a drop was always a **(19)** ____________________ of 1.6×10^{-19} C. Therefore, Millikan proposed that each **(20)** ____________________ always carried that charge.

In your textbook, read about objects sharing a charge.

Refer to the illustration. For each of the statements below, write true *or rewrite the italicized part to make the statement true.*

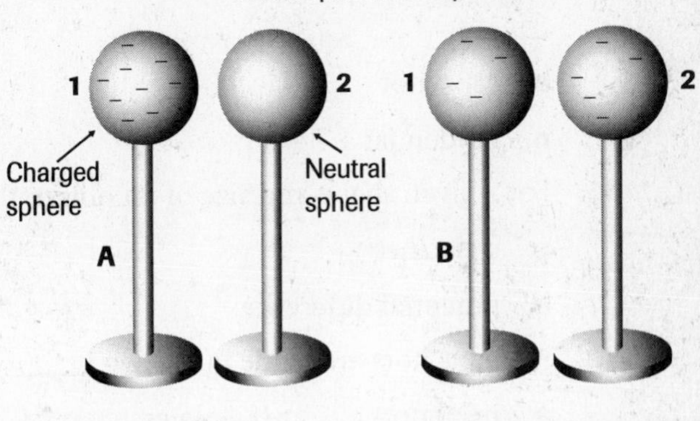

21. In drawing A, the electric potential of *sphere 1* is high.

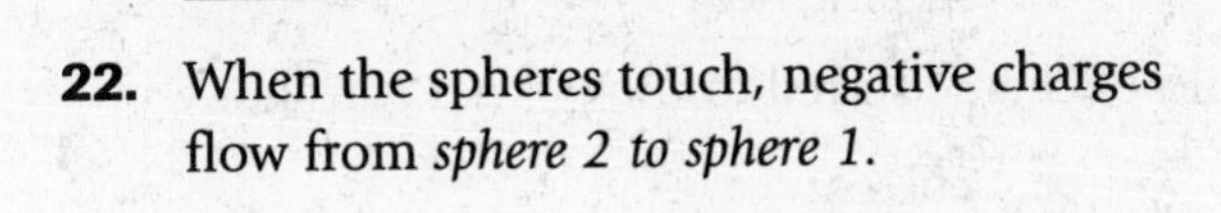

____

22. When the spheres touch, negative charges flow from *sphere 2 to sphere 1.*

____

23. When the spheres touch, the potential of sphere 2 *decreases* and the potential of sphere 1 decreases.

____

24. In drawing B, the potential of sphere 1 now *equals* the potential of sphere 2.

____

25. If sphere 2 were smaller than sphere 1, the two spheres *would not* have reached the same potential by touching.

____

Name ____________________

21 Study Guide

In your textbook, read about grounding and conductors.
Answer the following questions.

26. Why can an object be touched to Earth to eliminate excess charge on a body?

____

____

27. Give one example of an object that requires grounding to prevent damage or injury.

____

____

____

28. Explain why people inside a car are protected from electric fields generated by lightning.

____

29. Explain how a lightning rod works.

____

____

____

____

In your textbook, read about capacitance and capacitors.
Circle the letter of the answer that best completes each statement.

30. A small device invented by Pieter Van Musschenbroek that can store a large electric charge is the _____.

a. capacitor **c.** resistor
b. Leyden jar **d.** electric field

31. For a given shape and size of an object, the ratio of charge stored to _____ is a constant.

a. resistance **c.** current
b. potential difference **d.** size

32. All capacitors are made up of two _____, separated by an insulator.

a. insulators **b.** plates **c.** conductors **d.** wires

33. Capacitors are used in electric circuits to store _____.

a. charge **b.** current **c.** resistance **d.** capacitance

34. The capacitance of a capacitor is _____ the charge on it.

a. dependent on **b.** linked to **c.** changed by **d.** independent from

Date ______________ Period ______________ Name ______________________

22 Study Guide

Use with Chapter 22.

Current Electricity

Vocabulary Review

For each description on the left, write the letter of the matching item.

_____ **1.** a closed loop though which charges can flow
_____ **2.** the flow of positive charge
_____ **3.** a variable resistor or rheostat
_____ **4.** a flow of charged particles
_____ **5.** the property that determines how much current is present
_____ **6.** a flow of 1 C/s
_____ **7.** a circuit diagram
_____ **8.** a device that measures current
_____ **9.** several voltaic cells connected together
_____ **10.** device designed to have a specific resistance
_____ **11.** a connection that provides only one path for a current
_____ **12.** a device that measures the potential difference of a circuit
_____ **13.** a connection of two or more electric devices that provides more than one current path
_____ **14.** a device that converts light energy to electric energy
_____ **15.** 100 watts of electric energy delivered continuously and used for 1 h

a. ammeter
b. ampere
c. battery
d. conventional current
e. electric circuit
f. electric current
g. kilowatt-hour
h. parallel connection
i. photovoltaic cell
j. potentiometer
k. resistance
l. resistor
m. schematic
n. series connection
o. voltmeter

Name ______________________

22 Study Guide

Section 22.1: Current and Circuits

In your textbook, read about electric circuits.

Circle the letter of the choice that best completes the statement or answers the question.

1. If two conductors at different potential differences are connected by another conductor, charges flow from the conductor with the _____ potential difference to the conductor with the _____ potential difference.

   a. higher, lower **b.** lower, higher **c.** negative, positive. **d.** medium, higher

2. If two conductors, A and B, are connected by another conductor, when does the flow of charge stop?

   a. when the potential difference of A is lower than the potential difference of B

   b. when the potential difference of A is higher than the potential difference of B

   c. when the potential difference of all conductors is equal

   d. never

3. Batteries and generators are both sources of _____.

   a. light **c.** photovoltaic cells

   b. electric energy **d.** kinetic energy

4. A(n) _____ is a closed loop that consists of a charge pump connected to a device that reduces the potential energy of the flowing charges.

   a. electric circuit **c.** resistor

   b. generator **d.** potentiometer

5. The potential energy lost by the charges moving through a device in a circuit is represented by _____.

   a. ΔV **c.** PE

   b. E **d.** qV

6. A general term that could refer to a battery, photovoltaic cell, or generator in a circuit is _____.

   a. ammeter **c.** charge pump

   b. resistor **d.** current

7. No generator is 100 percent efficient, and the kinetic energy that is not converted to electric energy usually _____.

   a. stays electric energy **c.** becomes light energy

   b. becomes thermal energy **d.** disappears

8. Because q is conserved, the net change in potential energy of the charges going completely around the circuit _____.

   a. must be zero

   b. is higher at the beginning than at the end

   c. is lower at the beginning than at the end

   d. is a negative number

Name ______________________________

22 Study Guide

In your textbook, read about electric power.

Write the term that correctly completes each statement. Use each term once.

ammeter	**coulomb**	**energy**	**power**
ampere	**electric current**	**potential difference**	**watts**
charge			

Power measures the rate at which **(9)** ____________________ is transferred. The energy carried by an electric current depends on the **(10)** ____________________ transferred and the potential difference across which it moves. The unit used for quantity of electric charge is the **(11)** ____________________. Thus, the rate of flow of electric charge, or **(12)** ____________________, is measured in coulombs per second. One coulomb per second is a(n) **(13)** ____________________. A(n) **(14)** ____________________ is used to measure current. The **(15)** ____________________ of an electric device is found by multiplying the **(16)** ____________________ by the current. The power, or energy delivered to the electric device per second, is measured in joules per second, or **(17)** ____________________.

In your textbook, read about resistance.

For each of the statements below, write true *or rewrite the italicized part to make the statement true.*

18. ____________________ If you put a glass rod between two conductors that have a potential difference between them, you will have *a very large current.*

19. ____________________ *Resistance* is measured by placing a potential difference across two points on a conductor and measuring the current.

20. ____________________ Resistance is the ratio of the potential difference to the *charge.*

21. ____________________ The resistance of a conductor is measured in *ohms.*

22. ____________________ One ohm, 1Ω, is the resistance that permits a current of 1 A when a *potential difference of 1 V* is applied across the resistance.

23. ____________________ A device that has a *changing resistance* and appears to be independent of the potential difference is said to obey Ohm's law.

24. ____________________ To reduce current, you need a large resistance in a *large volume.*

25. ____________________ Superconductors are materials that have *high* resistance.

Name ______________________________

22 Study Guide

In your textbook, read about control of current.
Answer the following questions.

26. How does voltage affect the current that passes through a resistor?

____

____

27. How can the current be reduced if the voltage is kept constant?

____

____

28. Explain how a variable resistor, or potentiometer, works.

____

____

____

____

In your textbook, read about symbols used in schematic diagrams.
Complete the table below by writing the correct symbols or the names of components.

Component	Symbol	Component	Symbol
capacitor			(symbol: two crossing lines with a dot at the junction)
conductor		no electric connection	
fuse			(symbol: open switch)
lamp			

Name ____________________

22 Study Guide

In your textbook, read about schematic diagrams of circuits.
Complete the schematic diagrams below.

29. Label the components of the schematic diagram below.

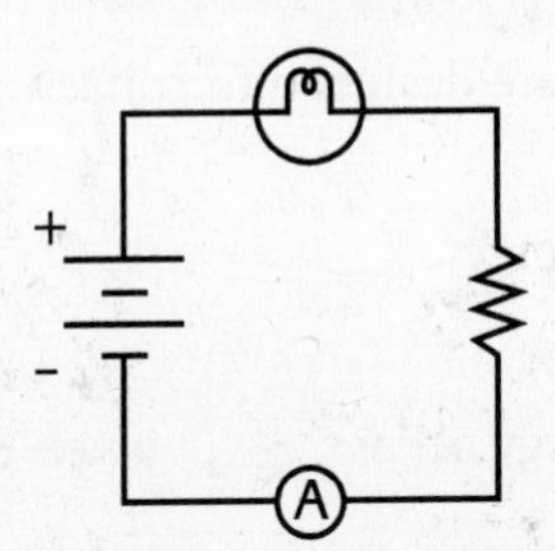

30. Draw a schematic diagram for the circuit in the drawing below.

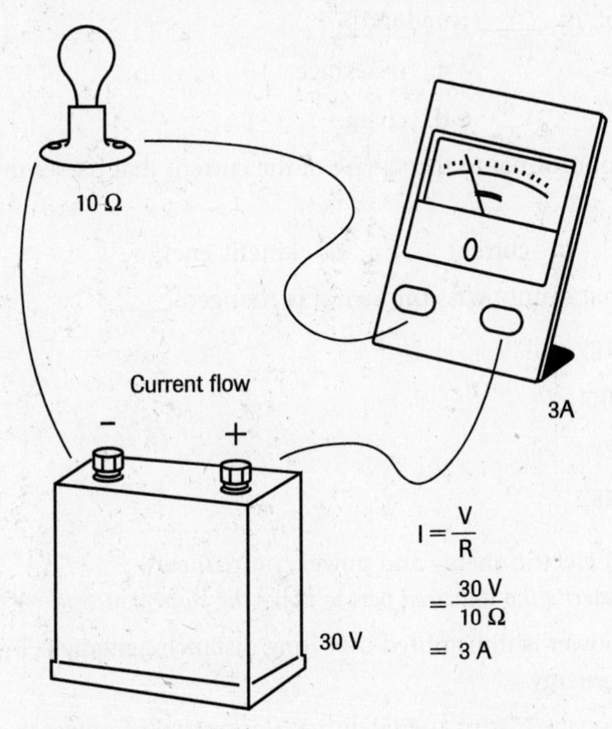

31. Draw a schematic diagram for a circuit that has a battery, a lamp, a motor, a switch, and an ammeter, with a voltmeter across the motor.

Name ______________________

22 Study Guide

Section 22.2: Using Electrical Energy

In your textbook, read about resistors used to turn electric energy into thermal energy.
Circle the letter of the choice that best completes each statement.

1. A space heater, a hot plate, and the heating element in a hair dryer are designed to convert almost all the electric energy into _____.

a. kinetic energy
b. potential energy
c. thermal energy
d. solar energy

2. Household appliances that convert electric energy into thermal energy act as _____ when they are in a circuit.

a. resistors **b.** motors **c.** ammeters **d.** voltage dividers

3. When a charge moves through a resistor, its _____ is reduced.

a. current
b. potential difference
c. resistance
d. charge

4. The _____ dissipated in a resistor is proportional to the square of the current that passes through it and to the resistance.

a. voltage **b.** power **c.** current **d.** kinetic energy

5. A resistor gets hot because the power that cannot pass through it is changed _____.

a. from thermal energy to electric energy
b. from kinetic energy to thermal energy
c. from electric energy to kinetic energy
d. from electric energy to thermal energy

In your textbook, read about transmission of electric energy and power companies.
For each of the statements below, write true *or rewrite the italicized part to make the statement true.*

6. ______________________ When power is transmitted over long distances, energy is lost as *thermal* energy.

7. ______________________ To reduce the loss of energy during transmittal of power over long distances, either the current or the *voltage* must be reduced.

8. ______________________ All *wires* have some resistance, even though the resistance is small.

9. ______________________ Cables of high conductivity and *small* diameter are used to transmit power long distances to reduce the resistance and reduce loss of energy as thermal energy.

10. ______________________ Current can be reduced without reducing power, by increasing the *voltage.*

11. ______________________ Electric companies measure their energy sales in *joules per second times seconds.*

Date ______________ Period ______________ Name ______________________

23 Study Guide

Use with Chapter 23.

Series and Parallel Circuits

Vocabulary Review

For each description on the left, write the letter of the matching item.

_____ **1.** circuit in which all current travels though each device

_____ **2.** short piece of metal that melts if too large a current passes through it

_____ **3.** occurs when a circuit forms that has a very low resistance

_____ **4.** circuit in which there are several different paths for a current

_____ **5.** automatic switch that opens a circuit when the current reaches some set value

_____ **6.** circuit that has some resistors in parallel and some in series

_____ **7.** value of a single resistor that could replace all resistors in a circuit without changing the current

_____ **8.** device used to measure the current in part of a circuit

_____ **9.** device used to measure the potential drop across some part of a circuit

_____ **10.** detects small differences in current caused by an extra current path and opens the circuit

_____ **11.** series circuit used to produce a voltage source from a higher-voltage battery

a. ammeter
b. circuit breaker
c. combination series-parallel circuit
d. equivalent resistance
e. fuse
f. ground-fault interrupter
g. parallel circuit
h. series circuit
i. short circuit
j. voltage divider
k. voltmeter

Name ______________________

23 Study Guide

Section 23.1: Simple Circuits

In your textbook, read about current in series circuits.

Circle the letter of the choice that best completes the statement or answers the question.

1. The current is _____ a series circuit.

a. higher at the beginning of

b. the same everywhere in

c. lower at the beginning of

d. variable in

2. In an electric circuit, the increase in voltage provided by the generator or other energy source, ΔV_{source}, is equal to the _____ of voltage drops across the resistors.

a. subtraction **b.** multiplication **c.** sum **d.** average

3. Which of the following equations is not correct?

a. $I = \frac{\Delta V_{source}}{(R_1 + R_2)}$

b. $I = \frac{\Delta V_{source}}{R}$

c. $I = \frac{\Delta V_{source}}{(R_1 + R_2 + R_3)}$

d. $I = R_3 + \frac{\Delta V_{source}}{(R_1 + R_2)}$

4. Which of the following equations correctly computes the equivalent resistance for a series circuit with four resistors?

a. $R = R_1 + R_2 + R_3 + R_4$

b. $R = R_1 \times R_2 \times R_3 \times R_4$

c. $\frac{1}{R} = \frac{1}{R_1} + \frac{1}{R_2} + \frac{1}{R_3} + \frac{1}{R_4}$

d. $R = \frac{(R_1 \times R_2)}{(R_3 \times R_4)}$

5. In a series circuit, the equivalent resistance is _____ any single resistance.

a. larger than

b. determined by

c. equal to

d. smaller than

6. If the battery voltage does not change, adding more devices in series _____ the current.

a. sometimes decreases

b. always decreases

c. sometimes increases

d. always increases

7. Given the voltage of a series circuit, you first calculate _____ to find the current through the circuit.

a. the voltage

b. the equivalent resistance

c. the power

d. the equivalent voltage

Name ______________________

23 Study Guide

In your textbook, read about voltage drops in series circuits.

Answer the following questions, using complete sentences.

8. Why must the net change in potential be zero as current moves through a circuit?

9. How do you find the potential drop across an individual resistor?

10. What type of circuit is used in a voltage divider?

11. What is the purpose of a voltage divider?

12. What determines the resistance of a photoresistor?

13. Why are photoresistors often used in voltage dividers? In what special devices can they be used?

Name ______________________

23 Study Guide

In your textbook, read about parallel circuits.

Refer to the circuit diagram below. Circle the letter of the best answer for each question.

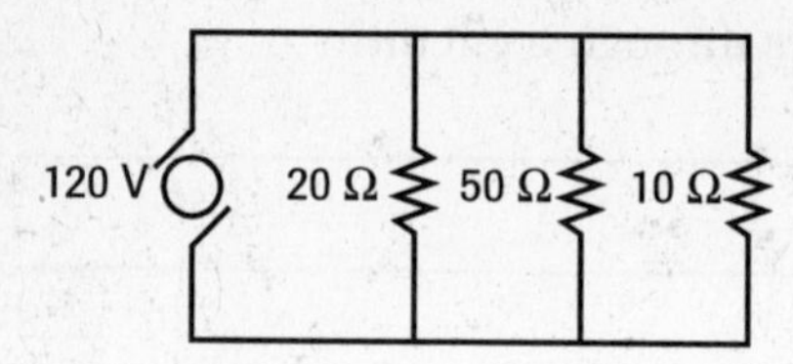

14. What type of circuit does this diagram represent?

a. series circuit

b. parallel circuit

c. combination series-parallel circuit

d. tandem circuit

15. How many current paths are in this circuit?

a. one

b. four

c. three

d. five

16. How would you calculate the total current of this circuit?

a. Total current is found by computing the average of the currents through each path.

b. Total current is found by adding the currents through each path.

c. Total current is found by subtracting the currents through each path.

d. Total current cannot be calculated for this circuit.

17. If the 10-Ω resistor were removed from the circuit, which of the following would *not* be true?

a. The current through the 20-Ω resistor would be unchanged.

b. The sum of the current in the branches of the circuit would change.

c. The total current through the generator would change.

d. The current through the 50-Ω resistor would change.

18. Which of the following is true for this circuit?

a. The equivalent resistance of this circuit is smaller than 10 Ω.

b. $R = R_1 + R_2 + R_3$

c. $R = \frac{1}{R_1} + \frac{1}{R_2} + \frac{1}{R_3}$

d. $R = R_1 \times R_2 \times R_3$

Name ______________________

23 Study Guide

Section 23.2: Applications of Circuits

In your textbook, read about safety devices.

For each of the statements below, write true *or rewrite the italicized part to make it true.*

1. When appliances are connected in *parallel,* each additional appliance placed in operation reduces the equivalent resistance in the circuit and causes more current to flow through the wires.

____

2. The *length* of the metal in a fuse determines the amount of current that will melt the fuse and break the circuit.

____

3. When a circuit breaker *opens,* it allows current to flow.

____

4. Ground-fault interrupters can be used as safety devices on circuits in which the current flows along *a single path* from the power source into the electric outlet and back to the source.

____

5. Electric wiring in homes uses only *series* circuits.

____

6. Low resistance causes the current to be very *small* and may result in a short circuit.

____

In your textbook, read about combined series-parallel circuits.

Circle the letter of the choice that best answers the question.

7. Which diagram represents a combined series-parallel circuit in which a 30-Ω resistor and a 75-Ω resistor are connected in parallel to a 125-V source through a 2-Ω resistor in series?

a.

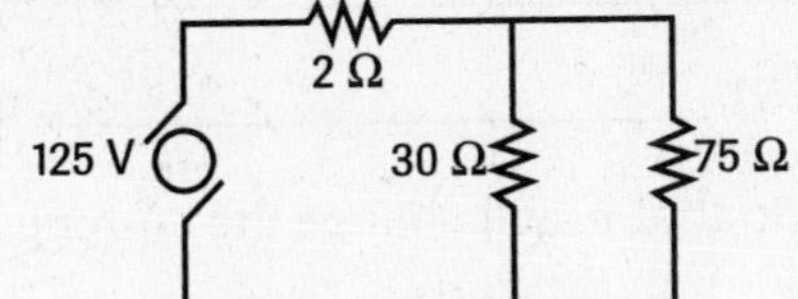

c.

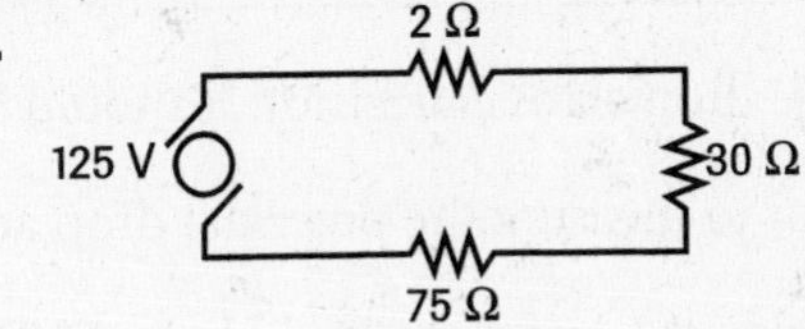

b.

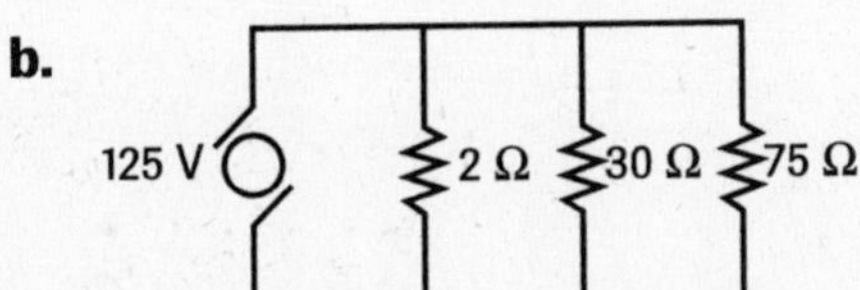

d.

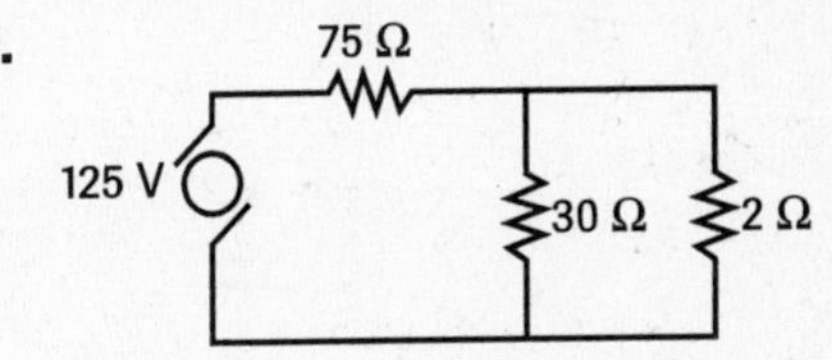

Name ______________________

23 Study Guide

In your textbook, read about ammeters and voltmeters.
Redraw the circuit diagram below according to the following directions.

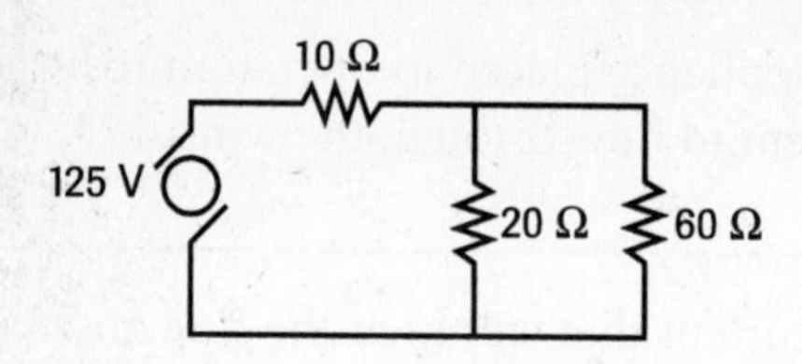

8. Insert an ammeter in the circuit that would measure the current of the entire circuit.

9. Insert an ammeter in the circuit that would measure the current that flows through the 60-Ω resistor.

10. Insert a voltmeter that would measure the voltage drop across the 10-Ω resistor.

Write the term that correctly completes each statement. Use each term once.

ammeter	**low**	**series**
high	**parallel**	**voltmeter**

A(n) **(11)** ______________________ measures current. It is placed in **(12)** ______________________ with the resistor if you want to measure the current through a resistor. So that it will change the current as little as possible, its resistance should be as **(13)** ______________________ as possible. A(n) **(14)** ______________________ measures the voltage drop across a resistor. It should be connected in **(15)** ______________________ with a resistor to measure the potential drop across that resistor. So that it will change the current as little as possible, its resistance should be as **(16)** ______________________ as possible.

Date ______________ Period ______________ Name ______________________

24 Study Guide

Use with Chapter 24.

Magnetic Fields

Vocabulary Review

For each description on the left, write the letter of the matching item.

_____ **1.** a current-carrying coil of wire that has a north pole and a south pole

_____ **2.** the strength of a magnetic field

_____ **3.** used to find the direction of the magnetic field around a current-carrying wire

_____ **4.** the combined magnetic fields of electrons in a group of atoms

_____ **5.** used to find the direction of the magnetic field around an electromagnet

_____ **6.** used to find the direction of the force on a current-carrying wire in a magnetic field

_____ **7.** a device used to measure very small currents

_____ **8.** the number of magnetic field lines passing through a surface

_____ **9.** having a north pole and a south pole

_____ **10.** the magnetic forces that exist around magnets

_____ **11.** imaginary lines used to help visualize a magnetic field

_____ **12.** a long coil of wire consisting of many loops

_____ **13.** a device that converts electrical energy to kinetic energy

_____ **14.** several rotating loops of wire in an electric motor

a. armature
b. domain
c. electric motor
d. electromagnet
e. first right-hand rule
f. galvanometer
g. magnetic field
h. magnetic field lines
i. magnetic flux
j. magnetic induction
k. polarized
l. second right-hand rule
m. solenoid
n. third right-hand rule

Name ______________________

24 Study Guide

Section 24.1 Magnets: Permanent and Temporary

In your textbook, read about magnets.

For each of the statements below, write true *or rewrite the italicized part to make it true.*

1. ______________________ When a magnet is allowed to swing freely, it comes to rest lined up in *an east-west* direction.

2. ______________________ A *compass* is a small magnet mounted so that it is free to turn.

3. ______________________ Like poles *attract* each other.

4. ______________________ Magnets *do not always* have two opposite poles.

5. ______________________ If Earth is considered a giant magnet, the south pole of the Earth-magnet is near Earth's *geographic north pole.*

6. ______________________ Many permanent magnets are made of *pure iron.*

In your textbook, read about magnetic fields.

Answer the following questions, using complete sentences.

7. Describe one method for showing the magnetic field around a magnet.

____

____

____

8. What does *magnetic flux* mean in terms of both magnetic field lines and magnetic field strength? Where is it greatest on a magnet?

____

____

____

9. Describe the direction and shape of magnetic field lines of a bar magnet.

____

____

____

____

10. Describe three things that happen when a sample made of iron, cobalt, or nickel is placed in the magnetic field of a permanent magnet.

____

____

____

____

Name ______________________

24 Study Guide

In your textbook, read about magnetic fields around current-carrying wires and electromagnets.
Answer the following questions, using complete sentences.

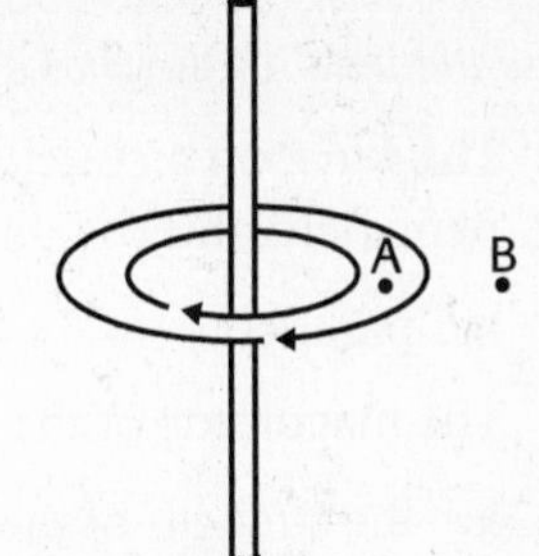

11. In the drawing at the right, what is the direction of the conventional current?

____

12. Is the strength of the magnetic field greater at point A or point B?

____

____

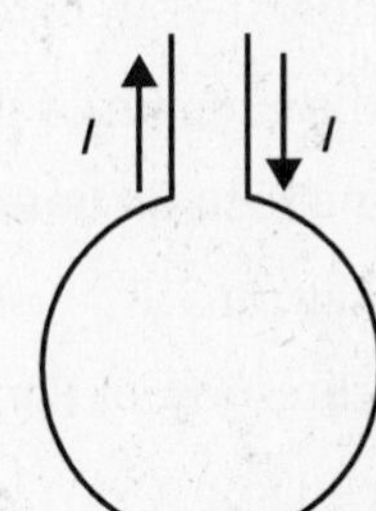
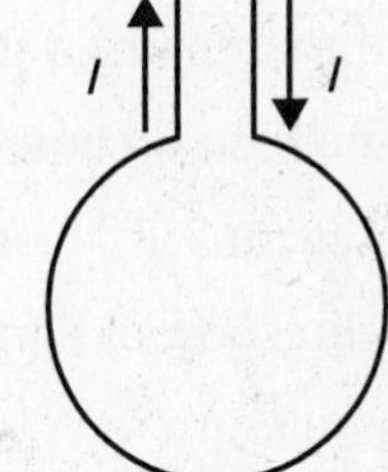

13. Describe the direction of the magnetic field inside and outside of the loop, shown at the right.

____

____

14. In the drawing of an electromagnet shown below, which end is the magnetic north pole?

____

____

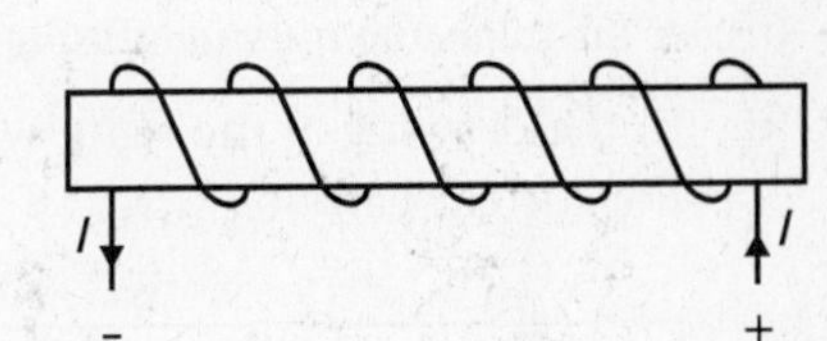

15. Describe two ways to increase the strength of the magnetic field around an electromagnet, besides placing an iron rod inside the coil.

____

____

____

In your textbook, read about domains in magnetic materials.
For each of the statements below, write true *or* false.

__________ **16.** The magnetic fields of the electrons in a group of neighboring atoms cannot combine together.

__________ **17.** When a piece of iron is not in a magnetic field, the domains point in random directions.

__________ **18.** In permanent magnets, the domains point in random directions.

__________ **19.** The material on the tape for a recorder is chosen so that the alignment of the domains is only temporary.

__________ **20.** The direction of the magnetization in different rocks on the seafloor varies, indicating that the north and south poles of Earth have exchanged places many times.

Name ______________________________

24 Study Guide

Section 24.2: Forces Caused by Magnetic Fields

In your textbook, read about forces on current-carrying wires in magnetic fields.
Circle the letter of the choice that best completes each statement.

1. The force on a current-carrying wire in a magnetic field is _____ both the direction of the magnetic field and the direction of current.

   a. parallel to **b.** at right angles to **c.** opposite **d.** independent of

2. The magnitude of the force on a current-carrying wire in a magnetic field is proportional to _____.

   a. the strength of the field, the current in the wire, and the length of the wire in the magnetic field

   b. only the strength of the field

   c. only the strength of the field and the current in the wire

   d. the strength of the field, the current in the wire, and the voltage in the wire

3. Magnetic induction is measured in _____.

   a. newtons **b.** teslas **c.** amperes **d.** volts

4. The direction of Earth's magnetic field is toward the _____.

   a. equator **c.** south magnetic pole

   b. north magnetic pole **d.** surface

In your textbook, read how to find the direction of force on a current-carrying wire in a magnetic field.
Answer the following questions, using complete sentences.

5. In the drawing at the right, what is the direction of the force on the current-carrying wire?

   ____

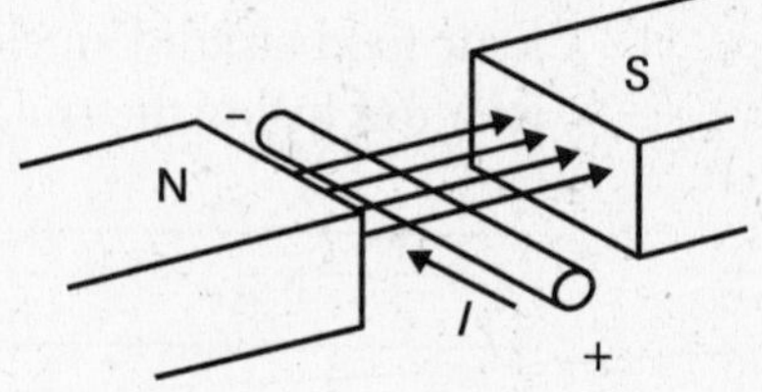

6. In the drawing at the right, what is the direction of the force on the current-carrying wire?

   ____

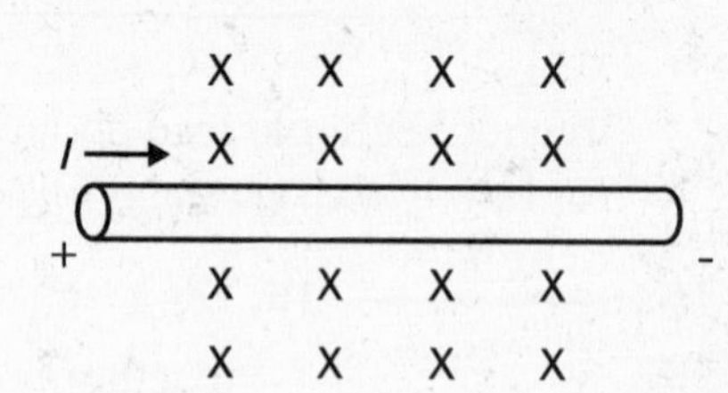

7. Are the current-carrying wires in the drawing at the right attracted or repelled?

   ____

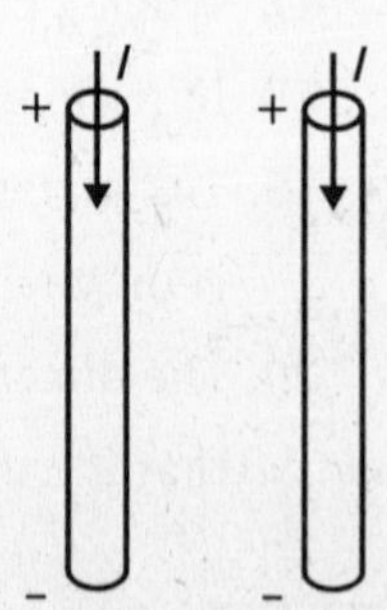

Name ______________________________

24 Study Guide

In your textbook, read about uses of the force on a current-carrying wire in a magnetic field.
For each of the following statements, write true *or* false.

__________ **8.** A loudspeaker changes sound energy into electrical energy.

__________ **9.** The amplifier driving the loudspeaker sends a current through a coil of wire mounted on a paper cone and placed in a magnetic field.

__________ **10.** The force exerted on the coil of wire in a magnetic field in a loudspeaker pushes the coil into or out of the field, depending on the magnitude of the current.

__________ **11.** The forces exerted on a current-carrying loop of wire in a magnetic field cannot be used to measure current.

__________ **12.** In an electric motor, the loop of current-carrying wire must rotate 180°.

In your textbook, read about galvanometers.
Write the term that correctly completes each statement. Use each term once.

current	**loop**	**proportional**	**torque**
direction	**magnetic fields**	**series**	**voltmeter**
down	**multiplier**	**spring**	**wire**
galvanometers	**parallel**		

(13) ______________________ that exert forces on a loop of wire carrying a current can be used to measure very small currents. A small loop of **(14)** ______________________ that is carrying a current is placed in the strong magnetic field of a permanent magnet. The current passing through the **(15)** ______________________ goes in one end and out the other. The **(16)** ______________________ of the force on the wire resulting from the magnetic field can be determined by using the third right-hand rule. One side of the loop is forced **(17)** ______________________; the other side of the loop is forced up. The resulting **(18)** ______________________ rotates the loop. The magnitude of the torque acting on the loop is proportional to the magnitude of the **(19)** ______________________. This principle is used in **(20)** ______________________ to measure small currents. A small spring in a galvanometer exerts a torque that opposes the torque resulting from the current; thus the amount of torque is **(21)** ______________________ to the current. A galvanometer can be used as a(n) **(22)** ______________________ or an ammeter. A galvanometer can be converted to an ammeter by connecting it in **(23)** ______________________ with a resistor having a resistance smaller than that of the galvanometer. To convert the galvanometer to a voltmeter, the galvanometer is connected in **(24)** ______________________ with a resistor called a **(25)** ______________________.

Name ____________________

24 Study Guide

In your textbook, read about force exerted on charged particles by magnetic fields.
Answer the following questions, using complete sentences.

26. Explain how pictures form on the screen of a cathode-ray tube in a computer monitor or television set.

27. On what does the force produced by a magnetic field on a single electron depend?

28. Why is the direction of the force on an electron in a magnetic field opposite to the direction given by the third right-hand rule?

29. Why does a charged particle in a uniform magnetic field have a circular path?

30. What do high-energy nuclear-particle accelerators use magnets for, and how does this process work?

31. Explain what causes the aurora borealis.

Date ______________ Period ______________ Name ______________________

25 Study Guide

Use with Chapter 25.

Electromagnetic Induction

Vocabulary Review

Circle the letter of the choice that best completes each statement.

1. The induction of *EMF* in a wire carrying changing current is _____.
a. mutual inductance
b. magnetic induction
c. electromotive force
d. self-inductance

2. The part of a transformer that is connected to a source of AC voltage is the _____.
a. armature **b.** primary coil **c.** secondary coil **d.** core

3. The potential difference given to charges by a charge pump is the _____.
a. electromotive force **b.** current **c.** electronic force **d.** eddy current

4. The law that states that the direction of the induced current is such that the magnetic field resulting from the induced current opposes the change in the field that caused the induced current is _____.
a. Ohm's law **b.** Newton's law **c.** Faraday's law **d.** Lenz's law

5. Generating a current in a circuit by relative motion between a wire and a magnetic field is _____.
a. electromagnetic induction
b. magnetism
c. electric induction
d. current induction

6. A device that increases the voltage is a _____.
a. step-down transformer
b. step-up transformer
c. capacitor
d. generator

7. A varying *EMF* is induced in the _____ of a transformer.
a. primary coil **b.** secondary coil **c.** tertiary coil **d.** armature

8. A(n) _____ increases or decreases AC voltages.
a. generator **b.** armature **c.** transformer **d.** electric motor

9. The induction of an *EMF* in one coil as a result of a varying magnetic field in another coil is _____.
a. self-inductance
b. electric induction
c. mutual inductance
d. magnetic induction

10. A(n) _____ converts mechanical energy to electric energy.
a. electric generator **b.** electric motor **c.** transformer **d.** capacitor

11. A(n) _____ is used to decrease voltage.
a. step-up transformer
b. step-down transformer
c. electric generator
d. electric motor

12. When a piece of metal moves through a magnetic field, _____ are produced. They produce a magnetic field that opposes the motion that caused them.
a. resistances **b.** electromagnets **c.** permanent magnets **d.** eddy currents

Name ______________________________

25 Study Guide

Section 25.1: Creating Electric Current from Changing Magnetic Fields

In your textbook, read about production of an electric current with a magnetic field.
Circle the letter of the choice that best completes each statement.

1. Michael Faraday found that _____ could be produced by moving a wire through a magnetic field.

a. a resistance **b.** a current **c.** charge **d.** magnetism

2. When a wire is held stationary or moved _____ to a magnetic field, no current flows.

a. parallel **b.** perpendicular **c.** tangent

3. An electric current is generated in a wire in a magnetic field only when _____.

a. the wire is stationary in the field

b. the wire cuts magnetic field lines

c. the wire is moved parallel to the field

d. a small current exists in the wire before it is placed in the magnetic field

4. To generate a current in a wire in a magnetic field, _____.

a. the conductor must move through the magnetic field while the magnetic field remains stationary

b. the magnetic field must move past the conductor while the conductor remains stationary

c. there must be relative movement between the wire and the magnetic field

d. there must be a battery connected to the wire

5. Electromotive force is not a force; it is a _____.

a. current **b.** resistance **c.** charge **d.** potential difference

6. *EMF* is measured in _____.

a. volts **b.** amperes **c.** ohms **d.** newtons

7. The *EMF* produced in a wire moving in a magnetic field depends only on the _____.

a. field strength

b. current in the wire

c. magnetic field strength, the length of the wire in the field, and the velocity of the wire in the field

d. magnetic field strength, the length of the wire in the field, and the current in the wire

8. If a wire moves through a magnetic field at an angle to the field, only the component of the wire's velocity that is _____ generates *EMF*.

a. positive

b. parallel to the direction of the magnetic field

c. negative

d. perpendicular to the direction of the magnetic field

Name ______________________

25 Study Guide

In your textbook, read about microphones and generators.
Answer the following questions, using complete sentences.

9. Explain how *EMF* is produced in a microphone.

____

____

____

10. Explain how a sound wave is converted to an electric signal in a microphone.

____

11. What is the purpose of an electric generator?

____

12. Describe how an electric generator works.

____

____

13. What is one way to increase the induced *EMF* produced by a generator?

____

14. Explain why the strength of the electric current changes as the armature of an electric generator turns.

____

____

____

____

15. Explain why the direction of the electric current changes as the armature of an electric generator turns.

____

____

____

16. Compare and contrast electric generators and electric motors.

____

____

____

____

____

Name ___________________

25 Study Guide

In your textbook, read how to find the direction of a current induced in a wire by a changing magnetic field. *Answer the following questions.*

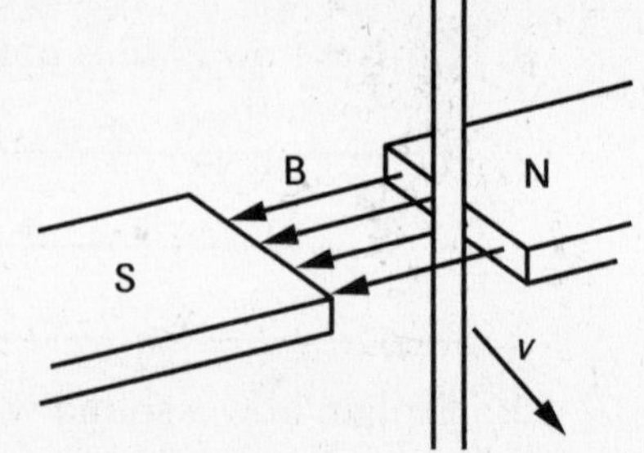

17. In the drawing on the right, in which direction does the current move?

____

18. Which right-hand rule is used to find the direction of the current in the drawing on the right?

____

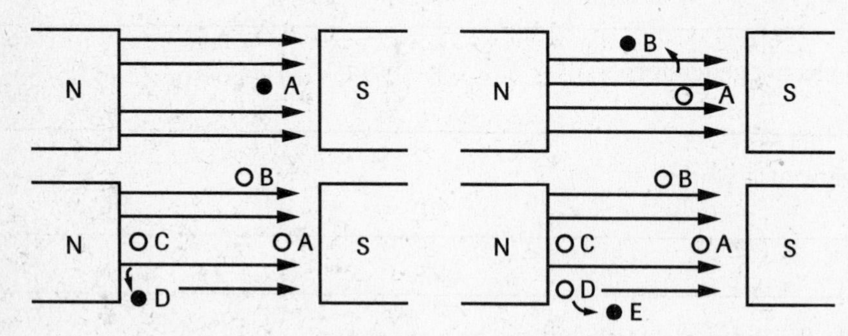

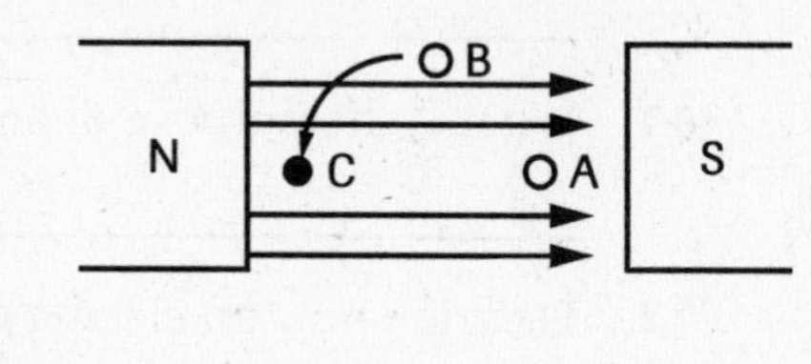

19. The drawing above represents five positions of one side of a wire loop as it turns in a magnetic field. The current produced when the wire loop is in position A is indicated on the graph below. Mark the approximate points on the graph that correspond to positions B through E.

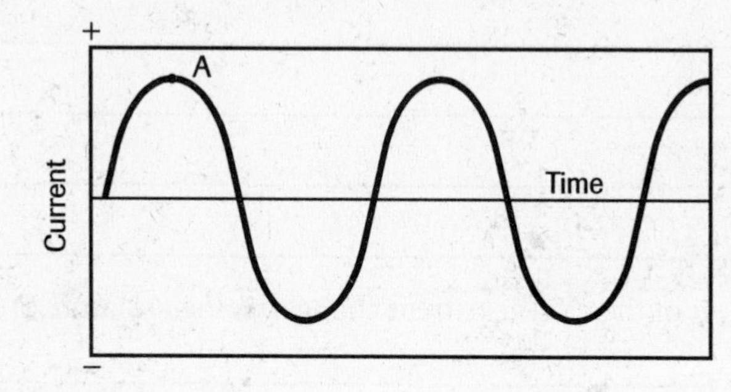

In your textbook, read about alternating-current generators.
For each of the statements below, write true *or* rewrite *the italicized part to make the statement true.*

20. _______________ In the United States, electric utilities use generators with a *60-Hz* frequency.

21. _______________ As the armature in a generator turns, the current *remains constant.*

22. _______________ The power associated with an alternating current is *constant.*

23. _______________ The power of an alternating current is always *positive.*

24. _______________ Alternating current and voltage are often described in terms of their *effective* current and voltage.

Name ______________________________

25 Study Guide

Section 25.2: Changing Magnetic Fields Induce *EMF*

In your textbook, read about Lenz's law.

For each of the statements below, write true *or rewrite the italicized part to make the statement true.*

1. ______________________ When a current is produced in a wire by electromagnetic induction, the direction of the current is such that the magnetic field produces a force on the wire that *opposes* the original motion of the wire.

2. ______________________ The direction of an induced current is such that the magnetic field resulting from the induced current opposes *the field* that caused the induced current.

3. ______________________ If the N-pole of a magnet is moved toward the right end of the coil, the right end of the coil *attracts* the magnet and becomes an N-pole.

4. ______________________ If a generator produces a larger current, then the armature will be *more difficult* to turn.

5. ______________________ The opposing force on an armature means that mechanical energy must be supplied to the generator to produce electric energy; this fact is consistent with *the law of conservation of energy.*

In your textbook, read about applications of Lenz's law.

Answer the following questions, using complete sentences.

6. Explain why a large current flows when an electric motor is first turned on but is reduced as the motor begins to turn.

____

____

____

____

7. What can cause a spark when a motor's plug is pulled from an outlet?

____

____

____

8. How does a sensitive balance use Lenz's law to stop its oscillation?

____

____

____

____

Name ______________________________

25 Study Guide

In your textbook, read about self-inductance.
Write the term that correctly completes each statement. Use each term once.

constant	*EMF*	magnetic field lines
current	increases	self-inductance
decreases	larger	zero

(9) ______________________ through a coil of wire generates a magnetic field. If the current increases, the magnetic field **(10)** ______________________. This can be pictured as creation of new **(11)** ______________________. As they expand, they cut through the coil of wires, generating a(n) **(12)** ______________________ to oppose the current increase. This generating process is called **(13)** ______________________. The faster the current changes, the **(14)** ______________________ the opposing *EMF* and the slower the current change. If the current reaches a steady value, the magnetic field is **(15)** ______________________, and the *EMF* is **(16)** ______________________. When the current **(17)** ______________________, the *EMF* generated helps prevent the reduction in magnetic field and current.

In your textbook, read about transformers.
Circle the letter of the choice that best completes each statement.

18. Transformers can change _____ with relatively little loss of energy.

a. power **b.** resistances **c.** voltages **d.** magnetic fields

19. A transformer has _____ coils, electrically insulated from each other, but wound around the same iron core.

a. two **b.** three **c.** four **d.** five

20. When the primary coil is connected to a source of AC voltage, the changing current creates a _____.

a. steady magnetic field
b. potential difference
c. varying magnetic field
d. resistance

21. In the secondary coil, the varying magnetic field induces a _____.

a. varying *EMF* **b.** steady *EMF* **c.** resistance **d.** charge

22. In a step-up transformer, the number of coils of wire in the primary coil is _____ the number of coils in the secondary coil.

a. equal to **b.** less than **c.** double **d.** greater than

23. In an ideal transformer, the electric power delivered to the secondary circuit is _____ the power supplied to the primary circuit.

a. equal to **b.** less than **c.** double **d.** greater than

Date ______________ Period ______________ Name ______________________

26 Study Guide

Use with Chapter 26.

Electromagnetism

Vocabulary Review

Write the letter of the choice that best completes each statement.

1. A wire that is connected to an alternating current source that forms or receives electromagnetic waves is a(n) _____.

a. receiver **b.** dish **c.** antenna **d.** cable

2. The combination of an antenna, coil and capacitor circuit, and amplifier is a _____.

a. receiver **c.** mass spectrometer

b. radio **d.** secondary antenna

3. Atoms that have the same chemical properties but different masses are _____.

a. nuclides **b.** isotopes **c.** isomers **d.** elements

4. Energy that is carried through space in the form of electromagnetic waves is _____.

a. electromagnetic radiation **c.** magnetic fields

b. electric fields **d.** electricity

5. Crystals that bend or deform when a voltage is applied across them show the property of _____.

a. conductance **c.** magnetism

b. electromagnetic waves **d.** piezoelectricity

6. An instrument that is used to find the masses of positive ions is a(n) _____.

a. mass spectrometer **c.** analytical scale

b. precise balance **d.** scintillation counter

7. Electromagnetic waves of relatively high frequency that are generated by accelerating electrons to high speeds using 20 000 or more volts and then crashing the electrons into matter are _____.

a. gamma rays **c.** X rays

b. infrared rays **d.** visible light rays

8. Electric and magnetic fields that move through space make up a(n) _____.

a. electric current **c.** static charge

b. electromagnetic wave **d.** electric wave

9. The type(s) of radiation with the longest wavelengths is(are) _____.

a. microwaves **c.** radio waves

b. light **d.** X rays and gamma rays

10. The type(s) of radiation with the highest frequency is(are) _____.

a. ultraviolet **c.** radio waves

b. light **d.** X rays and gamma rays

Name ______________________

26 Study Guide

Section 26.1: Interaction Between Electric and Magnetic Fields and Matter

In your textbook, read about Thomson's experiment to measure the mass of an electron.

Refer to the diagram below. Write the term or letter that correctly completes each statement. Use each item once.

cathode-ray tube	**electric field**	**magnitude**
charge-to-mass ratio	**electrons**	**straight path**
circular path	**forces**	**velocity**
direction	**magnetic field**	

The **(1)** ______________________ of an electron was first measured in 1897 by J. J. Thomson. For his experiment, Thomson used a(n) **(2)** ______________________ with all of the air removed. An electric field accelerated **(3)** ______________________ off the negatively charged cathode, labeled **(4)** ______________________ in the diagram, toward the positively charged anode, labeled **(5)** ______________________. Electrons that passed through a hole in the anode followed a(n) **(6)** ______________________ toward a fluorescent screen, which glowed where the electrons hit.

Electric and magnetic fields in the center of the tube exerted **(7)** ______________________ on the electrons. The **(8)** ______________________ was perpendicular to the beam of electrons. The **(9)** ______________________ was at right angles to both the beam and the electric field. When the electric field was stronger, the electrons struck the fluorescent screen at the point labeled **(10)** ______________________. When the magnetic field was stronger, the electrons struck the fluorescent screen at point **(11)** ______________________. When the electric and magnetic fields were adjusted until the forces of the two fields were equal in **(12)** ______________________ and opposite in **(13)** ______________________, the electrons struck the screen at point **(14)** ______________________. The magnetic and electric field forces were balanced only for electrons that had a specific **(15)** ______________________. If the electric field was turned off, only the force due to the magnetic field remained, and the electrons followed a(n) **(16)** ______________________.

Name ____________________

26 Study Guide

In your textbook, read about Thomson's experiment to measure the mass of a proton.
For each of the statements below, write true *or* false.

__________ **17.** Positively charged particles, when they are in an electric field, bend the same direction as electrons.

__________ **18.** Positively charged particles, when they are in an magnetic field, bend the opposite way from electrons.

__________ **19.** When electrons are pulled off hydrogen atoms, the hydrogen has a net negative charge.

__________ **20.** The mass of the proton was determined from the charge-to-mass ratio.

__________ **21.** The mass of the proton was determined to be 9.107×10^{-31} kg.

In your textbook, read about mass spectrometers.
Circle the letter of the choice that best completes each statement.

22. When Thomson put neon gas into the tube, he found two dots on the screen instead of one, because neon has two _____.

a. atoms **b.** isomers **c.** isotopes **d.** masses

23. The masses of positive ions can be measured precisely using a(n) _____, an adaptation of a cathode-ray tube.

a. ordinary balance **c.** mass spectrometer
b. analytic scale **d.** ion accelerator

24. Accelerated _____ strike the ion source of a mass spectrometer, knocking off electrons and thus forming positive gas ions.

a. electrons **b.** protons **c.** ions **d.** atoms

25. In a mass spectrometer, a(n) _____ produces an electric field to accelerate the ions.

a. magnetic field **c.** antenna
b. potential difference **d.** magnet

26. The ions that are undeflected move into a region with a uniform magnetic field where they follow a(n) _____.

a. narrow path **b.** straight path **c.** circular path **d.** elliptical path

27. The charge-to-mass ratio of the ion is determined by the equation _____.

a. $q/m = 2V/B^2r^2$ **b.** $r = mv/qB$ **c.** $KE = mv^2/2$ **d.** $q/m = 2V/Br$

28. In the mass spectrometer, the ions leave a mark when they hit a _____.

a. fluorescent screen **c.** glass screen
b. cathode-ray tube **d.** photographic film

29. Besides being used to measure the mass of positive ions, mass spectrometers can also _____.

a. make negative ions from atoms **c.** separate isotopes of atoms
b. combine atoms **d.** create isotopes of atoms

Name ______________________________

26 Study Guide

Section 26.2: Electric and Magnetic Fields in Space

In your textbook, read about electromagnetic waves.

For each of the statements below, write true *or rewrite the italicized part to make the statement true.*

1. ______________________ A changing magnetic field can produce a changing electric field *without* a wire.

2. ______________________ Maxwell postulated that a changing electric field produces a *constant* magnetic field.

3. ______________________ Either accelerating charges or changing magnetic fields can produce *combined electric and magnetic fields* that move through space, known as electromagnetic waves.

4. ______________________ Electromagnetic waves move at the speed of *sound.*

5. ______________________ In order of *increasing* wavelength, the types of radiation are radio waves, microwaves, infrared, light, ultraviolet, and X rays and gamma rays.

In your textbook, read about production of electromagnetic waves.

Circle the letter of the choice that best completes each statement.

6. The alternating current source connected to an antenna produces a _____.

a. constant current that produces a constant electric field

b. constant current that produces a changing electric field

c. changing current that produces a constant electric field

d. changing current that produces a changing electric field

7. The changing electric field produced by an antenna generates a _____.

a. wind

b. constant magnetic field

c. changing magnetic field

d. charge

8. The magnetic field of an electromagnetic wave oscillates _____ the electric field.

a. parallel to

b. at lower frequency than

c. at higher frequency than

d. at right angles to

9. The electric field and magnetic field of an electromagnetic wave are _____ the direction of the motion of the wave.

a. at right angles to

b. opposite to

c. parallel to

d. independent of

10. An electromagnetic wave produced by an antenna is _____.

a. energetic

b. out of phase

c. faster than the speed of light

d. polarized

Name ____________________

26 Study Guide

In your textbook, read about using a coil and a capacitor to generate high-frequency electromagnetic waves.

For each of the statements below, write true *or rewrite the italicized part to make the statement true.*

11. ____________________ When the battery is removed from a circuit containing a coil and capacitor, the *coil* discharges and the stored electrons flow through the coil, creating a magnetic field.

12. ____________________ When the magnetic field of the coil collapses, a *back-EMF* recharges the capacitor in the opposite direction.

13. ____________________ To increase the oscillation frequency, the size of the coil and the capacitor must be made *larger.*

14. ____________________ Microwaves and infrared waves *cannot* be generated using a coil and capacitor and an antenna.

In your textbook, read about the pendulum analogy for the coil and capacitor circuit.

For each pendulum analogy, write the letter of the matching coil and capacitor property.

_____ **15.** the pendulum bob

_____ **16.** The pendulum bob moves fastest when its displacement from vertical is zero.

_____ **17.** When the bob is at its greatest angle, it has zero velocity.

_____ **18.** The potential energy of the pendulum is largest when its displacement is greatest. The kinetic energy is largest when the velocity is greatest.

_____ **19.** The sum of the potential and kinetic energies and thermal energy losses of the pendulum is constant.

_____ **20.** The pendulum will eventually stop swinging if it is left alone.

_____ **21.** Gentle pushes applied at the correct times will keep a pendulum moving.

a. The largest current flows through the coil when the charge on the capacitor is zero.

b. When the current is largest, the energy stored in the magnetic field is greatest. When the current is zero, the electric field of the capacitor is largest.

c. Voltage pulses applied to the coil and capacitor circuit at the right frequency will keep the oscillations going.

d. the electrons in the coil and capacitor

e. When the capacitor holds the largest charge, the current through the coil is zero.

f. Oscillations of the coil and capacitor will die out if energy is not added to the circuit.

g. The sum of the magnetic field energy, the electric field energy, the thermal losses, and the energy carried away by electromagnetic fields is constant.

Name ______________________

26 Study Guide

In your textbook, read about detection of electromagnetic waves.

Answer the following questions, using complete sentences.

22. Explain how electromagnetic waves are detected by an antenna.

____

____

23. An antenna designed to receive radio waves is much longer than one designed to receive microwaves. Why is there a size difference?

____

____

____

24. Describe two ways in which the detected *EMF* can be increased.

____

____

____

____

25. Explain how radio and television waves of a certain frequency can be selected. Describe briefly how selection works.

____

____

____

In your textbook, read about X rays.

Circle the letter of the choice that best completes each statement.

26. If electrons have _____ kinetic energy with a very _____ voltage, X rays are produced when the electrons are stopped suddenly by matter.

a. low, high   **c.** low, low

b. high, high   **d.** high, low

27. X rays can penetrate _____, but are blocked by _____.

a. lead, bone   **c.** bone, lead

b. bone, soft body tissue   **d.** soft body tissue, bone

28. When electrons crash into matter, _____ are converted into X rays.

a. the electrons   **c.** their potential energies

b. their kinetic energies   **d.** the atoms in the matter

Date ______________ Period ______________ Name ______________________

27 Study Guide

Use with Chapter 27.

Quantum Theory

Vocabulary Review

Circle the letter of the choice that best completes each statement.

1. For light to cause the emission of an electron of a specific metal, its frequency must be at least the _____.
   - **a.** threshold frequency.
   - **b.** work function
   - **c.** de Broglie frequency
   - **d.** emission frequency
2. Discrete bundles of energy of light and other forms of radiation are _____.
   - **a.** quarks
   - **b.** photons
   - **c.** electromagnetic waves
   - **d.** isotopes
3. The emission of electrons when electromagnetic radiation falls on an object is the _____.
   - **a.** threshold frequency
   - **b.** Heisenberg principle
   - **c.** Compton effect
   - **d.** photoelectric effect
4. _____ states that the position and momentum of a particle cannot be precisely known at the same time.
   - **a.** The Heisenberg uncertainty principle
   - **b.** The third law of motion
   - **c.** The position uncertainty law
   - **d.** Compton's law
5. The energy needed to free the most weakly bound electron from a metal is the _____ of the metal.
   - **a.** threshold frequency
   - **b.** minimum energy
   - **c.** work function
   - **d.** weak electron energy
6. The wavelength of a particle is referred to as its _____.
   - **a.** photon
   - **b.** de Broglie wavelength
   - **c.** threshold wavelength
   - **d.** Compton effect
7. The increase in wavelength of X-ray photons when they are scattered off electrons is the _____.
   - **a.** Compton effect
   - **b.** Heisenberg uncertainty principle
   - **c.** X scatter effect
   - **d.** photon principle
8. Energy that exists only in packages of specific amounts is _____.
   - **a.** specified. **b.** pulsing. **c.** quantized. **d.** discontinuous.
9. A body that is so hot that it glows is _____.
   - **a.** incandescent **b.** fluorescent **c.** phosphorescent **d.** radioactive

Name ______________________________

27 Study Guide

Section 27.1: Waves Behave Like Particles

In your textbook, read about radiation from incandescent bodies.

For each of the following statements, write true *or rewrite the italicized part to make the statement true.*

1. ______________________ Hot bodies contain *charges,* which radiate electromagnetic waves.
2. ______________________ Light and infrared radiation are produced by the vibration of particles within *an incandescent body.*
3. ______________________ Energy is emitted from incandescent bodies at *one specific frequency* that depends on the temperature.
4. ______________________ As the temperature of a hot body increases, the frequency at which the maximum energy is emitted *decreases.*
5. ______________________ The total power emitted by an incandescent body *increases* with increasing temperature.

In your textbook, read about Planck's theory that energy is quantized.

Circle the letter of the choice that best completes the statement or answers the question.

6. Max Planck calculated the spectrum of radiation emitted from incandescent bodies based on his assumption that ______.
   - **a.** energy is a wave
   - **b.** energy is not continuous
   - **c.** matter is a wave
   - **d.** matter is a particle
7. In the equation $E = nhf$, which of the following is *not* a possible value of n?
   - **a.** 0
   - **b.** 3
   - **c.** 5/4
   - **d.** 6
8. Planck's theory depends on the fact that ______.
   - **a.** atoms always radiate electromagnetic waves
   - **b.** atoms always radiate electromagnetic waves when they are vibrating
   - **c.** atoms emit radiation only when their vibration energy changes
   - **d.** atoms emit radiation only if they have a source of electricity
9. If the energy of an atom changes from an initial energy to a final energy, the energy radiated is equal to ______.
   - **a.** the initial energy
   - **b.** the final energy
   - **c.** the product of the initial energy and the final energy
   - **d.** the change in energy
10. Because the constant h is extremely ______, the energy-changing steps are too ______ to be noticeable in ordinary bodies.
    - **a.** small, small
    - **b.** large, small
    - **c.** large, large
    - **d.** small, large

Name ______________________

27 Study Guide

In your textbook, read about the photoelectric effect.

For each of the statements below, write true *or rewrite the italicized part to make the statement true.*

11. ______________________ The *absorption of energy* when electromagnetic radiation hits an object is called the photoelectric effect.

12. ______________________ Electromagnetic radiation of *any* frequency will eject electrons from a metal.

13. ______________________ If the frequency of the incident light shone on a metal is below the threshold frequency, no level of *intensity* will cause ejection of electrons.

14. ______________________ The *electromagnetic wave theory* cannot explain why electrons are ejected immediately even in dim light if the frequency of the radiation is at or above the threshold level.

In your textbook, read about Einstein's theories on the photoelectric effect.

Answer the following questions, using complete sentences.

15. How does Einstein's photoelectric-effect theory explain the existence of a threshold frequency?

____

____

____

16. What does Einstein's theory say about conservation of energy?

____

____

____

17. Why will all of the photoelectrons ejected from the same metal not have the maximum kinetic energy?

____

____

____

18. Briefly describe an experiment that supports Einstein's theory for the photoelectric effect.

____

____

____

____

____

____

Name ______________________

27 Study Guide

In your textbook, read about the kinetic energy of photoelectrons.
Refer to the graphs below that show the maximum kinetic energy of photoelectrons versus frequency for two different metals. Then answer the questions.

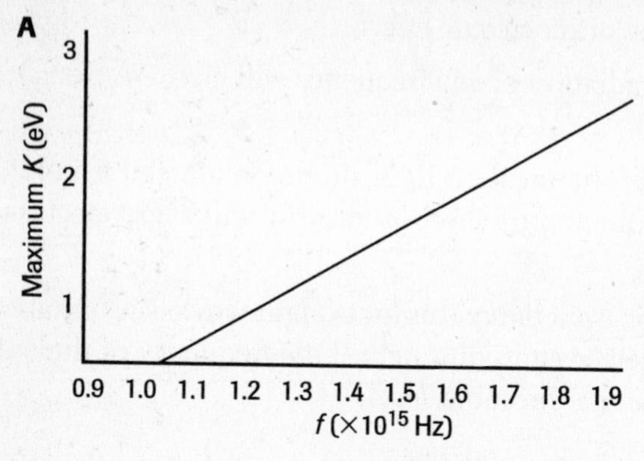

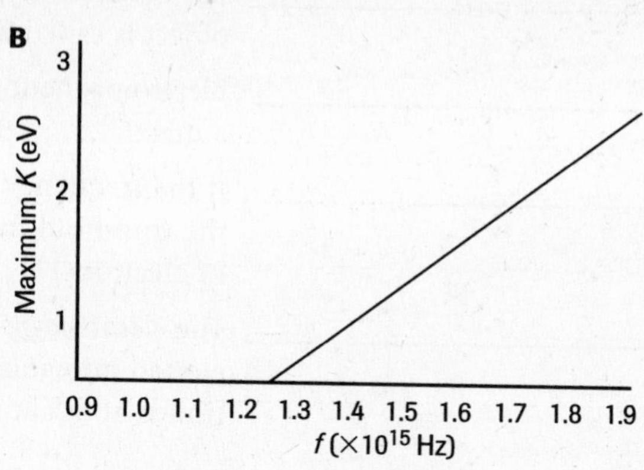

19. What is the numerical value of the slope of the line in graph A? In graph B?

____

20. To what constant are the slopes equal?

____

21. How can you tell that these are graphs for two different metals?

____

22. What is the approximate threshold frequency in graph A?

____

23. What is the minimum frequency required to produce photoelectrons from the metal in graph B?

____

24. At what frequency do the photoelectrons produced from the metal in graph A have no kinetic energy?

____

25. Will a frequency of 1.0×10^{15} Hz eject electrons from the metal in graph B? Why or why not?

____

26. What is the relationship between the frequency of the radiation and the kinetic energy of the photoelectrons?

____

____

27. How would you use the graphs to find the work function of these metals?

____

____

Name ______________________

27 Study Guide

In your textbook, read about the Compton effect.

Answer the following questions, using complete sentences.

28. Which of Einstein's theories did Compton's experiment test?

29. Briefly describe Compton's experiment.

30. What were the results of Compton's experiment?

31. How is wavelength related to the energy of a photon?

32. What did Compton's results mean?

33. In later experiments, Compton observed that electrons were ejected from the graphite block. What hypothesis did Compton suggest to explain this?

34. How did Compton test his hypothesis and what did he find?

35. How is a photon like matter?

36. How is a photon not like matter?

Name ______________________________

27 Study Guide

Section 27.2: Particles Behave Like Waves

In your textbook, read about the wave nature of matter.

Write the term that correctly completes each statement. Use each term once.

de Broglie wavelength	electrons	waves	$\lambda = h/p$
diffracted	objects	X rays	$p = h/\lambda$
diffraction grating	particles		

Louis-Victor de Broglie was first to suggest that particles could behave like **(1)** ______________________. He proposed that the momentum of a particle is represented by the equation **(2)** ______________________. Therefore, the wavelength of a particle is **(3)** ______________________. The wavelength of a particle is known as the **(4)** ______________________. One of the problems with de Broglie's theory was that effects such as diffraction and interference had never been observed for **(5)** ______________________. Thomson performed an experiment that showed that electrons are **(6)** ______________________ just as light is. In his experiment, he aimed a beam of **(7)** ______________________ at a very thin crystal. The crystal acted as a **(8)** ______________________. Electrons diffracted from the crystal formed the same pattern that **(9)** ______________________ of a similar wavelength formed. The wave nature of most **(10)** ______________________ is not observable because the wavelengths are extremely short.

In your textbook, read about the dual wave and particle description of light and matter.

Circle the letter of the choice that best completes each statement.

11. The spreading of light used to locate a particle can be reduced by ______ of the light.

- **a.** decreasing the wavelength
- **b.** decreasing the frequency
- **c.** increasing the wavelength
- **d.** increasing the intensity

12. The Compton effect means that, when light of short wavelengths strikes a particle, the ______ of the particle is changed.

- **a.** size
- **b.** momentum
- **c.** charge
- **d.** identity

13. According to the Heisenberg uncertainty principle, it is impossible to measure precisely both the ______ and the ______ of a particle at the same time.

- **a.** charge, position
- **b.** position, momentum
- **c.** charge, mass
- **d.** wavelength, momentum

Date ____________ Period ____________ Name ______________________

28 Study Guide

Use with Chapter 28.

The Atom

Vocabulary Review

Write the term that correctly completes each statement. Use each term once.

absorption spectrum	**excited state**	**quantum number**
***α* particle**	**ground state**	**scintillation**
coherent light	**incoherent light**	**spectroscope**
electron cloud	**laser**	**spectroscopy**
emission spectrum	**quantum mechanics**	**stimulated emission**
energy level	**quantum model**	

1. ______________________ A(n) _____ is a relatively massive, positively charged, fast-moving particle released by an atom.

2. ______________________ The lowest energy level for an electron is the _____.

3. ______________________ In _____, the light waves are in step.

4. ______________________ A small flash of light emitted by a detection screen is a(n) _____.

5. ______________________ A(n) _____ has dark lines in an otherwise continuous spectrum.

6. ______________________ A device that produces light by stimulated emission is a(n) _____.

7. ______________________ A(n) _____ is a quantized energy state of an electron in an atom.

8. ______________________ The _____ of an atom is the region in which there is a high probability of finding an electron.

9. ______________________ A field of physics that uses the quantum model of the atom to make predictions about atomic structure and behavior is _____.

10. ______________________ In _____, the light waves are not in step.

11. ______________________ A(n) _____ is the set of light wavelengths given off by an atom.

12. ______________________ The study of the light given off by atoms is _____.

13. ______________________ When atoms in an excited state are struck by other photons, they release photons in _____.

14. ______________________ The _____ predicts the probability that an electron is at a specific location.

15. ______________________ A device used to study the light given off by an atom is a(n) _____.

16. ______________________ The principal _____ determines the values of r and E.

17. ______________________ An electron that has absorbed energy is in a(n) _____.

Name ______________________

28 Study Guide

Section 28.1: The Bohr Model of the Atom

In your textbook, read about the nuclear model of the atom.

The diagram below shows an experiment, with a close-up of the atoms in area C. Refer to the diagram to answer the following questions.

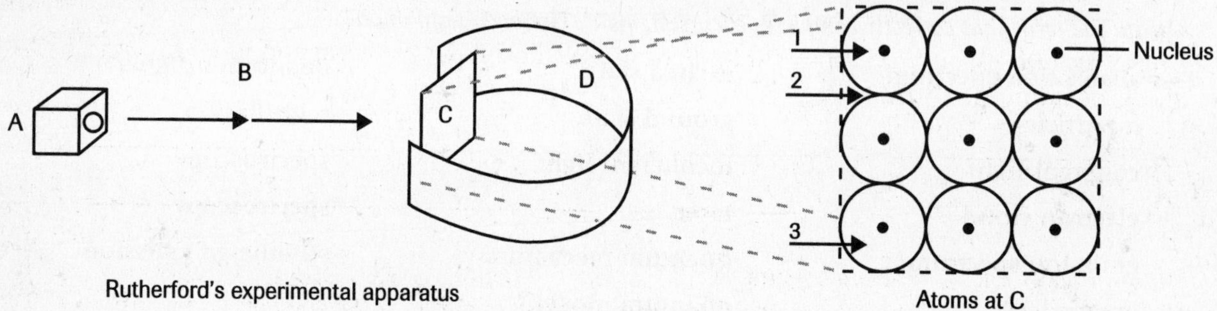

1. What kind of material does container A hold?

____

2. What makes up ray B? What characteristics does the ray have?

____

____

3. What type of material is C?

____

4. What is D, and what role does it play?

____

____

5. What will happen to the particle labeled 1? Why?

____

____

6. What will happen to the particle labeled 2? Why?

____

____

7. What will happen to the particle labeled 3? Why?

____

____

Name ______________________

28 Study Guide

8. Will most of the particles that pass into object C behave like particle 1, 2, or 3? What does that tell us about atoms?

____

____

9. From his experiment, what did Rutherford conclude about the atom and its component parts, including electrons?

____

____

In your textbook, read about spectra.

Circle the letter of the choice that best completes the statement or answers the question.

10. The spectrum of an incandescent solid is typically _____.
    - **a.** a set of colored lines
    - **b.** a single colored line
    - **c.** a band of colors with dark lines
    - **d.** a continuous band of colors

11. What information can an emission spectrum reveal about the elements present in a sample?
    - **a.** their identities only
    - **b.** their relative concentrations only
    - **c.** both their identities and their relative concentrations
    - **d.** neither their identities nor their relative concentrations

12. The Fraunhofer lines appear in the spectrum of _____.
    - **a.** sodium
    - **b.** all flames
    - **c.** the moon
    - **d.** the sun

13. Which element was discovered through study of the Fraunhofer lines?
    - **a.** hydrogen
    - **b.** helium
    - **c.** lithium
    - **d.** sodium

14. If the planetary theory of electrons were correct, at what wavelengths would light be released by an atom?
    - **a.** at all wavelengths
    - **b.** in a band at the red end only
    - **c.** at certain specific values
    - **d.** in a band at the violet end only

15. The visible spectrum of hydrogen consists of _____.
    - **a.** two lines
    - **b.** four lines
    - **c.** a single narrow band
    - **d.** a continuous wide band

16. The energy of a photon of light released by an atom is equal to _____.
    - **a.** hc/λ
    - **b.** $hc\lambda$
    - **c.** λ/f
    - **d.** $h\lambda/f$

Name ______________________

28 Study Guide

17. What does the fact that an emission spectrum contains only certain wavelengths show about electrons?

    a. Their energy is continuous.
    b. Their energy is quantized.
    c. They are present in the nucleus.
    d. They have a negative charge.

In your textbook, read about the Bohr model of atoms.

For each of the statements below, write true *or rewrite the italicized part to make the statement true.*

18. ______________ Bohr hypothesized that the laws of electromagnetism *do not* operate inside atoms.

19. ______________ An electron *releases* energy when it moves from the ground state to an excited state.

20. ______________ The orbits of electrons in excited states tend to be *smaller* than those of electrons in the ground state.

21. ______________ According to the Bohr model, the energy of an *electron* in an atom is quantized.

22. ______________ The electromagnetic radiation emitted by hydrogen when the electron drops into its ground state is *infrared waves.*

23. ______________ The electromagnetic radiation emitted by hydrogen when the electron drops into the second energy level from higher levels is *visible light.*

In your textbook, read about the symbols used in Bohr's calculations.

For each term on the left, write the matching item.

_____	24. charge	λ
_____	25. Coulomb's constant	E
_____	26. angular momentum	h
_____	27. radius of orbit	K
_____	28. principal quantum number	mvr
_____	29. Planck's constant	n
_____	30. wavelength	q
_____	31. total energy	r

Section 28.2: The Quantum Model of the Atom

In your textbook, read about the shortcomings of the Bohr model of the atom and about the quantum model.

Circle the letter of the choice that best completes the statement or answers the question.

1. Which quantity equals a whole-number multiple of the wavelength of an electron?

    a. r
    b. πr
    c. $2\pi r$
    d. πr^2

Name ______________________

28 Study Guide

2. The modern quantum model of the atoms precisely predicts _____ for an electron at a given moment.
   a. the probability of being at a specific location
   b. position and momentum
   c. position and velocity
   d. direction and position
3. The probability of finding an electron in the electron cloud of an atom is _____.
   a. zero
   b. low
   c. high
   d. infinite
4. A shortcoming of the Bohr model of the atom is its inability to _____.
   a. provide correct values for the lines in the hydrogen spectrum
   b. explain why angular momentum is quantized
   c. explain chemical properties of the elements
   d. predict the ionization energy for hydrogen
5. Application of the theories of electromagnetism to Bohr's model predicted _____.
   a. a positive charge for electrons
   b. collapse of the atom
   c. great stability for the atom
   d. a highly diffuse nucleus
6. Wavelength of a particle is equal to _____.
   a. h/c
   b. mv/h
   c. h/mv
   d. mvr
7. Which of the following is *not* true of quantum theory?
   a. It is based on the wave model.
   b. It predicts the probability that the electron is at a given radius.
   c. It proposes a planetary picture of the atom.
   d. It can be used to give details of the structure of molecules.
8. The circumference of the Bohr orbit equals _____.
   a. $2\pi r$
   b. πr^2
   c. λ
   d. $\lambda/2\pi r$
9. Which of the following is true of n?
   a. It is always a negative number.
   b. It cannot be expressed as a number.
   c. It may take on any value.
   d. It is always a whole number.

In your textbook, read about the structure and operation of lasers.
Write the term that correctly completes each statement.

Light emitted by a(n) **(10)** ______________________ source has many wavelengths and moves in all directions. Because the waves are out of step with each other, this light is **(11)** ______________________. When an atom in an excited state is struck by a(n) **(12)** ______________________ with the correct amount of energy, the atom will emit another **(13)** ______________________ of identical **(14)** ______________________, which will be in step with the first one. In a laser, a flash of light

28 Study Guide

that has a(n) **(15)** ______________________ wavelength than that of the laser is used to excite the atoms. As the atoms decay to a(n) **(16)** ______________________ energy state, they lase, or give off light by **(17)** ______________________ emission. As photons are produced, they strike other **(18)** ______________________, which give off more photons. The photons are trapped inside a glass **(19)** ______________________ capped at each end with parallel **(20)** ______________________, one of which is 100% **(21)** ______________________. The other is only partially **(22)** ______________________ and allows some of the photons through. Because all the photons produced by the laser are in step, the light is **(23)** ______________________. The beam produced is very narrow and thus very **(24)** ______________________. It also is all one wavelength, so it is **(25)** ______________________.

In your textbook, read about applications of laser technology.
For each of the statements below, write true *or* false.

__________ **26.** A hologram is a photographic recording of the phase and intensity of light.

__________ **27.** Lasers are used in spectroscopy to remove atoms.

__________ **28.** Fiber optics transmits lasers, using total internal reflection.

__________ **29.** Laser beams spread out greatly over long distances.

__________ **30.** Lasers can cut a wide variety of materials.

Date ______________ Period ______________ Name ______________________

29 Study Guide

Use with Chapter 29.

Solid State Electronics

Vocabulary Review

Write the term that correctly completes each statement. Use each term once.

band theory	**extrinsic semiconductor**	**junction transistor**
conduction band	**forbidden gap**	***n*-type semiconductor**
depletion layer	**forward-biased diode**	***p*-type semiconductor**
diode	**hole**	**transistor**
dopant	**intrinsic semiconductor**	**valence band**

1. ______________ The lowest band that is not filled to capacity with electrons is the _____.

2. ______________ A semiconductor that conducts by means of holes is a(n) _____.

3. ______________ A(n) _____ conducts as a result of thermally freed electrons and holes.

4. ______________ The highest band that contains electrons is the _____.

5. ______________ A(n) _____ has a region of one type of doped semiconductor sandwiched between layers of the opposite type.

6. ______________ A(n) _____ is produced by adding impurity atoms to a semiconductor.

7. ______________ An impurity added to a semiconductor is a(n) _____.

8. ______________ An empty energy level in the valence band is a(n) _____.

9. ______________ A(n) _____ consists of a *p*-type semiconductor joined with an *n*-type semiconductor.

10. ______________ The region around a diode junction that has neither holes nor free electrons is the _____.

11. ______________ Any simple switch made of doped semiconducting material is a(n) _____.

12. ______________ A semiconductor that conducts by means of electrons is a(n) _____.

13. ______________ A range of energy values that no atom can possess is a(n) _____.

14. ______________ The _____ of solids explains electric conduction in terms of energy bands and forbidden gaps.

Name ______________________________

29 Study Guide

Section 29.1: Conduction in Solids

In your textbook, read about the band theory of solids.

For each of the statements below, write true *or rewrite the italicized part to make the statement true.*

1. ______________________ Electric charges are easy to displace in *conductors.*
2. ______________________ When two atoms are brought close together, their electric fields *don't affect* one another.
3. ______________________ When a large number of atoms is brought close together, their energy levels spread into *narrow* bands.
4. ______________________ There are forbidden gaps of energy for atoms that are *far apart* in large numbers.
5. ______________________ The band theory explains electric conduction in *gases.*
6. ______________________ The *inner bands* of atoms in a solid are full.
7. ______________________ The lowest band that is not filled to capacity with electrons is the *valence* band.
8. ______________________ The highest band that contains electrons is the *conduction* band.
9. ______________________ In conductors, the valence band and the conduction band are the *same.*
10. ______________________ The size of the forbidden gap between a full valence band and the conduction band determines whether a solid is an *insulator.*

In your textbook, read about conductors and insulators.

Circle the letter of the choice that best completes the statement or answers the question.

11. _____ is a very good conductor of electricity.
    - **a.** Silicon
    - **b.** Aluminum
    - **c.** Sulfur
    - **d.** Iodine
12. A material that conducts electricity well tends to have _____.
    - **a.** no valence electrons
    - **b.** partially filled bands
    - **c.** completely filled bands
    - **d.** conduction and valence bands that are far apart

Name ______________________

29 Study Guide

13. When an electric field is applied on a length of wire, ____.
    a. there is random motion but no overall drifting in one direction
    b. there is overall drifting in one direction, but no random motion
    c. there are both random motion and overall drifting in one direction
    d. there is neither random motion nor overall drifting in one direction
14. How is conductivity related to resistivity?
    a. The two are directly related.
    b. The two are equal.
    c. Conductivity is the reciprocal of resistivity.
    d. Their relationship can't be described in general terms.
15. What happens to the conductivity of metals as temperature increases?
    a. It remains the same.
    b. It increases.
    c. It decreases.
    d. It follows no general rule.
16. ____ is an insulator.
    a. Copper
    b. Silicon
    c. Gold
    d. Table salt
17. In an insulating material, ____.
    a. the valence band is filled and the conduction band is empty
    b. the valence band is empty and the conduction band is filled
    c. both the valence band and the conduction band are empty
    d. both the valence band and the conduction band are filled
18. What is the effect if a relatively small electric field is placed across an insulator?
    a. No current flows.
    b. A small current flows.
    c. A large current flows.
    d. The insulator acts like a conductor.

In your textbook, read about semiconductors.
For each item on the left, write the letter of the matching item.

____ 19. a pure semiconductor
____ 20. an atom with four valence electrons
____ 21. any impurity added to a semiconductor
____ 22. an empty energy level in the valence band
____ 23. any semiconductor that conducts because of added impurities

a. dopant
b. extrinsic semiconductor
c. hole
d. intrinsic semiconductor
e. silicon

29 Study Guide

In your textbook, read about doped semiconductors.
Write the term that correctly completes each statement.

Dopants provide extra **(24)** ______________ or extra negatively charged **(25)** ______________ to a semiconductor. If an atom with five valence electrons replaces an atom with four valence electrons, one electron is unneeded for **(26)** ______________. The extra electron serves as a **(27)** ______________ electron. Adding a small amount of **(28)** ______________ to the electron will move it into the **(29)** ______________ band. The resulting conductor is a(n) **(30)** ______________ semiconductor; it carries electricity by means of particles that have a(n) **(31)** ______________ charge. If an atom with three valence electrons replaces an atom with four valence electrons, an extra **(32)** ______________ forms. The atoms can accept electrons and can be used as a(n) **(33)** ______________ semiconductor. In either type of semiconductor, the net charge on the semiconductor is **(34)** ______________. As temperature increases, the electric conductivity of semiconductors **(35)** ______________ and their resistivity **(36)** ______________. This also happens to the electric conductivity of certain semiconductors when **(37)** ______________ intensity upon them increases.

Section 29.2: Electronic Devices

In your textbook, read about diodes.
The diagrams below show a diode used with two kinds of biasing. Refer to the diagrams to answer the following questions.

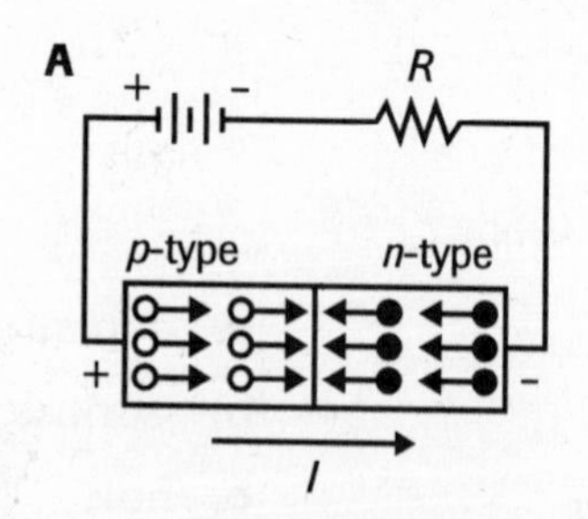

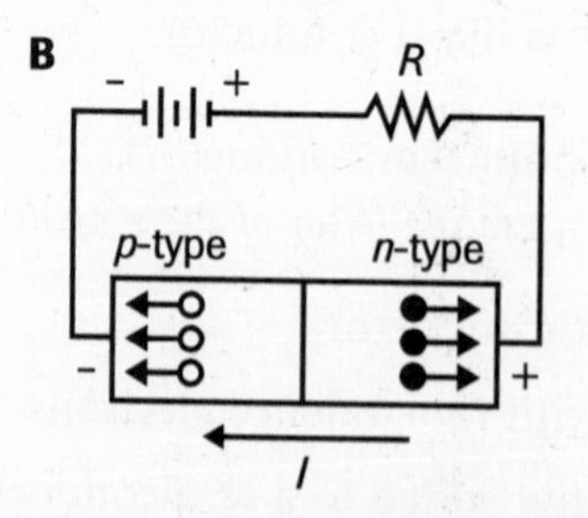

1. What function does the battery serve?

____

____

Name ______________________

29 Study Guide

2. Which way is conventional current flowing in each diagram, clockwise or counterclockwise?

____

3. Which way are electrons flowing in each diagram?

____

4. What differs in the *p*-type and *n*-type semiconductors?

____

____

5. What kind of biasing does each diode have? How you can tell?

____

____

____

6. Compare the depletion layer in the two diodes.

____

7. Compare the quantity of current flow through the two diodes. Which is a conductor, and which is a large resistor?

____

____

In your textbook, read about transistors.

Circle the letter of the choice that best completes the statement or answers the question.

8. Transistors are used mainly as .
   - **a.** resistors
   - **b.** voltage amplifiers
   - **c.** rectifiers
   - **d.** insulators

9. How many layers of semiconductor are in a simple junction transistor?
   - **a.** one
   - **b.** two
   - **c.** three
   - **d.** four

10. What type of transistor has a center that is an *n*-type semiconductor?
    - **a.** *n* transistor
    - **b.** *p* transistor
    - **c.** *npn* transistor
    - **d.** *pnp* transistor

29 Study Guide

11. In an *npn* transistor, conventional current flows from the _____.
    a. base to the emitter
    b. emitter to the base
    c. diode to the collector
    d. rectifier to the collector
12. In an *npn* transistor, electrons flow from the _____.
    a. emitter to the base to the collector
    b. collector to the base to the emitter
    c. base to the collector to the emitter
    d. base to the emitter to the collector
13. What carries current in a *pnp* transistor?
    a. electrons
    b. holes
    c. protons
    d. nothing
14. Current through the collector is _____ than current through the base.
    a. a little smaller
    b. a little larger
    c. much smaller
    d. much larger

Date ______________ Period ______________ Name ______________________

30 Study Guide

Use with Chapter 30.

The Nucleus

Vocabulary Review

Write the term that correctly completes each statement. Use each term once.

activity	gamma decay	nuclear reaction
alpha decay	gluon	nuclide
antiparticle	grand unification theory	pair production
atomic mass unit	graviton	positron
atomic number	half-life	quark
beta decay	lepton	strong nuclear force
elementary particle	mass number	transmute
force carrier	neutrino	weak boson

1. ______________ The nucleus of an isotope is a(n) ____.

2. ______________ The number of decays per second is the ____ of a radioactive substance.

3. ______________ During some types of radioactive decay, the radioactive element will ____, or change into another element.

4. ______________ A member of the family of fundamental particles that carry forces between matter is a ____.

5. ______________ The ____ family of fundamental particles includes protons, neutrons, and mesons.

6. ______________ The electron and neutrino are examples of the ____ family of fundamental particles.

7. ______________ A particle of antimatter is a(n) ____.

8. ______________ Energy can be converted to matter when a gamma ray passes close by a nucleus and ____ occurs.

9. ______________ Physicists are trying to create a ____ that connects the electromagnetic, weak, strong, and gravitational forces.

10. ______________ The particle that carries the strong force that holds the nucleus together is the ____.

11. ______________ The emission of an α particle is ____.

12. ______________ The little neutral particle emitted during beta decay is the ____.

13. ______________ The attractive force between protons and neutrons that are close together is the ____.

14. ______________ The emission of a β particle is ____.

Name ______________________________

30 Study Guide

15. ______________________ The particle involved in the weak nuclear interaction is the _____.
16. ______________________ The particle that carries the gravitational force is the _____.
17. ______________________ An antielectron, or _____, has the same mass as an electron but a positive charge.
18. ______________________ The number of protons in a nucleus is the atom's _____.
19. ______________________ The time required for one-half of a radioactive sample to decay is the _____ of that element.
20. ______________________ The sum of the number of protons and neutrons in a nucleus is the _____.
21. ______________________ A nucleus emits a gamma ray during the process of _____.
22. ______________________ Whenever the number of protons or neutrons in a nucleus changes, a(n) _____ occurs.
23. ______________________ The mass of one proton, 1.66×10^{-27} kg, is equal to one _____.
24. ______________________ A fundamental unit of matter that does not appear to be divisible into smaller units is a(n) _____.

Section 30.1: Radioactivity

In your textbook, read about the atomic nucleus.

For each of the statements below, write true *or* false.

__________ **1.** The only charged particle in the nucleus is the proton.

__________ **2.** Most of the mass of an atom comes from the mass of protons.

__________ **3.** The weak nuclear force overcomes the electrical force between protons to hold the nucleus together.

__________ **4.** Neutrons are not affected by the strong nuclear force.

__________ **5.** All nuclides of an element have the same number of protons but different numbers of neutrons.

__________ **6.** The mass of an individual atom is close to a whole number of mass units, while the atomic mass of an average sample of that type of atom is not, because it contains a mixture of isotopes.

__________ **7.** One mass unit is now defined on the basis of the mass of a proton.

__________ **8.** In the notation ${}^{A}_{Z}X$, A is the mass number, Z is the atomic number, and X is any element.

Name ______________________

30 Study Guide

In your textbook, read about radioactive decay.

Circle the letter of the choice that best completes each statement.

9. Photographic plates that Becquerel had near uranium compounds were exposed because _____.

a. their coverings had a light leak

b. uranium reacted chemically with the photographic plates

c. uranium emits light rays

d. uranium undergoes radioactive decay

10. Three types of radiation produced by radioactive decay are _____.

a. alpha, beta, and gamma

b. gamma, X ray, and light

c. ultraviolet, light, and infrared

d. alpha, beta, and delta

11. A thick sheet of paper can stop _____.

a. X rays **b.** gamma rays **c.** β particles **d.** alpha radiation

12. α particles are _____.

a. high-speed electrons

b. the nucleus of a helium atom

c. high-energy photons

d. low-energy photons

13. β particles are emitted when _____ during radioactive decay.

a. a proton changes to a neutron

b. a neutron changes to a proton

c. an electron changes energy levels

d. a valence electron is ejected

14. When a nucleus emits a gamma ray during gamma decay, _____.

a. both the mass number and the atomic number change

b. the atomic number increases by one and the mass number stays the same

c. neither the mass number nor the atomic number changes

d. the mass number decreases by four and the atomic number stays the same

In your textbook, read about nuclear reactions and equations.

For each of the statements below, write true *or rewrite the italicized part to make the statement true.*

15. ______________________ A *chemical reaction* occurs whenever the number of neutrons or protons in a nucleus changes.

16. ______________________ *All* nuclear reactions occur with a release of energy.

17. ______________________ Nuclear reactions can release excess energy in the form of *kinetic energy* of emitted particles.

18. ______________________ Nuclear reactions *cannot* be written in equation form.

19. ______________________ The total number of nuclear particles *stays the same* during a nuclear reaction.

20. ______________________ In a nuclear equation, the sum of the atomic numbers on the right side must *equal* the sum of the atomic numbers on the left side.

Name ______________________

30 Study Guide

In your textbook, read about the half-lives of radioactive elements.
Refer to the graph below to answer the following questions. Use complete sentences.

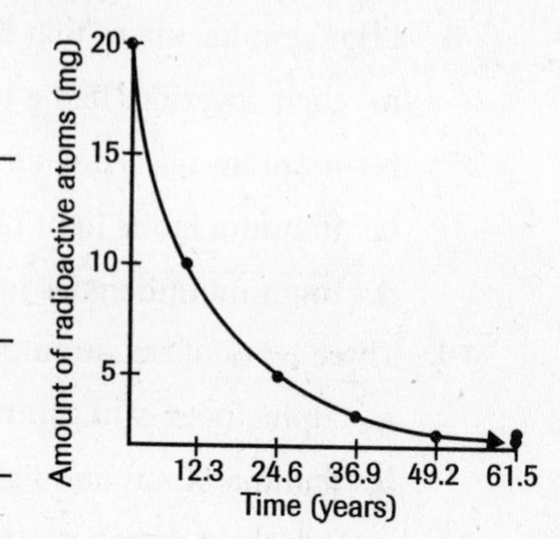

21. What fraction of the original amount of radioactive atoms remains after 24.6 years?

____

22. How many milligrams of radioactive atoms remain after 12.3 years?

____

23. Define half-life.

____

____

24. What is the half-life of the element in the graph?

____

25. At what time will the activity of the sample be one-eighth of the original activity?

____

26. Based on Table 30-1 in your textbook, what is the identity of the element represented in the graph?

____

27. According to Table 30-1, what type of radiation is produced by this element?

____

Section 30.2: The Building Blocks of Matter

In your textbook, read about nuclear bombardment and nuclear accelerators.
Answer the following questions, using complete sentences.

1. Why are neutrons often used to bombard nuclei?

____

2. Explain how a linear accelerator accelerates protons.

____

____

____

____

3. Why can synchrotrons be smaller than linear accelerators?

____

Name ______________________

30 Study Guide

In your textbook, read about particle detectors.

For each of the statements below, write true *or rewrite the italicized part to make the statement true.*

4. ______________________ *Exposure of photographic film* is one way to detect radiation from radioactive decay.

5. ______________________ The *Geiger-Mueller tube* detects radiation because high-speed particles emitted during radioactive decay ionize matter they bombard.

6. ______________________ In a *Wilson cloud chamber,* trails of small vapor droplets form in a liquid held just above the boiling point.

7. ______________________ Modern experiments use *spark chambers* in which a discharge is produced in the path of a particle passing through the chamber.

8. ______________________ Because *charged* particles do not produce charges and leave tracks, the laws of conservation and momentum in collisions are used to detect production of the particles in spark chambers.

In your textbook, read about elementary particles.

Complete the table by marking quark, lepton, or force carrier for each particle.

Table 1

	Particle	Quark	Lepton	Force Carrier
9.	electron			
10.	photon			
11.	neutrino			
12.	muon			
13.	charm			
14.	down			
15.	gluon			
16.	bottom			

In your textbook, read about particles and antiparticles.

Circle the letter of the choice that best completes each statement.

17. The wide range of energies of electrons emitted during beta decay suggested that another particle, later found to be the ——, was emitted with the β particle.

a. photon **b.** antineutrino **c.** electron **d.** muon

18. A proton within the nucleus can change into a neutron with the emission of a ______ and a neutrino.

a. positron **b.** antineutrino **c.** muon **d.** photon

Name ______________________________

30 Study Guide

19. The existence of beta decay indicates that there is a force, called the _____, acting in the nucleus.
    a. strong nuclear force
    b. beta force
    c. transient nuclear force
    d. weak nuclear force
20. When a particle and an antiparticle collide, the two can annihilate each other, resulting in _____.
    a. conversion of matter into energy
    b. creation of energy
    c. pair production
    d. creation of matter
21. Matter and antimatter particles must be produced in pairs to satisfy the law of conservation of _____.
    a. energy b. matter c. charge d. quarks

In your textbook, read about quarks and leptons.
Answer the following questions.

22. Which quarks make up a proton? A neutron? How do they differ?

    ____

    ____

23. Why can't individual quarks be observed?

    ____

24. Describe the steps that occur during beta decay using the quark model.

    ____

    ____

In your textbook, read about unification theories.
Write the term that correctly completes each statement.

The force between charged particles is carried by **(25)** ____________________. The electric force acts over a long distance because a photon has **(26)** ____________________. The weak force acts over short distances because a boson has **(27)** ____________________. In the high-energy collisions produced in accelerators, the **(28)** ____________________ and weak interactions have the same strength and range. Astrophysical theories indicate that, during massive stellar explosions, the two interactions are **(29)** ____________________. These theories unify the electromagnetic and weak forces into a single **(30)** ____________________. Physicists are trying to create a grand unification theory that includes the **(31)** ____________________ as well.

Date ______________ Period ______________ Name ______________________________

31 Study Guide

Use with Chapter 31.

Nuclear Applications

Vocabulary Review

For each term on the left, write the letter of the matching item.

_____ **1.** the part of a nuclear reactor that absorbs neutrons and controls the rate of the chain reaction

_____ **2.** the difference between the energy of the assembled nucleus and the particles in it

_____ **3.** the union of small nuclei to form larger ones

_____ **4.** liquefied fuel enclosed in tiny glass spheres and imploded by laser beams

_____ **5.** a nuclear reactor that produces more fuel than it uses

_____ **6.** the difference between the mass of the nucleus and the masses of the particles in it

_____ **7.** a proton or a neutron

_____ **8.** necessity for safe use of energy from fusion

_____ **9.** the substance in a nuclear reactor that slows fast neutrons

_____ **10.** a division of a nucleus into two or more fragments

_____ **11.** reactions that take place when nuclei have large amounts of thermal energy

_____ **12.** the process in which neutrons released by one fission induce other fissions

_____ **13.** a neutron released by a fission reaction and moving at a high speed

_____ **14.** a neutron that has been slowed by a moderator

_____ **15.** a process that increases the number of fissionable nuclei

a. binding energy
b. breeder reactor
c. chain reaction
d. control rod
e. controlled fusion
f. enrichment
g. fast neutron
h. fission
i. fusion
j. inertial confinement fusion
k. mass defect
l. moderator
m. nucleon
n. slow neutron
o. thermonuclear reactions

Name ______________________

31 Study Guide

Section 31.1: Radioactivity

In your textbook, read about what holds the nucleus together.

For each of the statements below, write true *or rewrite the italicized part to make the statement true.*

1. ______________________ The force that overcomes the mutual repulsion of the charged protons is the *electromagnetic force.*

2. ______________________ An assembled nucleus has *more* energy than the separate protons and neutrons in it.

3. ______________________ The binding energy comes from the nucleus's converting some of its *mass to energy.*

4. ______________________ Because mass is converted to energy to hold the nucleons together, the mass of an assembled nucleus is *more* than the sum of the masses of the nucleons that compose it.

5. ______________________ The mass defect is the *total* of the sum of the masses of the individual nucleons and the actual mass.

6. ______________________ The binding energy is calculated using the equation $E = mc^2$ to compute the energy equivalent of the experimentally determined mass defect.

In your textbook, read about how the nuclear binding energy relates to nuclear reactions.

Circle the letter of the choice that best completes each statement.

7. Nuclei whose atomic masses are larger than that of iron are _____ strongly bound and are therefore _____ stable than iron.

- **a.** more, less
- **b.** less, less
- **c.** more, more
- **d.** less, more

8. A nuclear reaction will occur naturally if energy is _____ by the reaction.

- **a.** released
- **b.** absorbed
- **c.** used up
- **d.** conserved

9. When a nucleus undergoes radioactive decay, the resulting nucleus is _____ the original nucleus.

- **a.** less stable than
- **b.** more stable than
- **c.** as stable as

10. At atomic numbers below 26, reactions that add nucleons to the nucleus make the nucleus _____.

- **a.** lighter
- **b.** more reactive
- **c.** more stable
- **d.** less stable

Name ______________________________

31 Study Guide

Section 31.2: Using Nuclear Energy

In your textbook, read about the production of artificial radioactivity.
Write the term that correctly completes each statement.

In 1934, Irene Joliot-Curie and Frederic Joliot bombarded aluminum with **(1)** ______________________, producing phosphorus atoms and neutrons. **(2)** ______________________ were also produced, even after the alpha bombardment stopped. The positrons came from the new **(3)** ______________________. Radioactive isotopes arise from stable isotopes bombarded with α particles, **(4)** ______________________, neutrons, electrons, or gamma rays. The resulting unstable nuclei may emit alpha radiation, **(5)** ______________________, gamma radiation, and positrons.

In your textbook, read about artificial radioactivity used in medicine.
Answer the following questions, using complete sentences.

6. Explain how tracer isotopes are used in medicine.

____

____

____

____

7. Name an instrument that can detect the decay products of tracer isotopes.

____

____

____

8. Explain how a PET scanner works.

____

____

____

____

____

9. How can $^{60}_{27}Co$ be used in medicine?

____

Name ____________________

31 Study Guide

In your textbook, read about nuclear fission.

Complete each statement.

10. The division of a nucleus into two or more fragments is ____________________.

11. In a fission reaction, the total mass of the products is ____________________ than the total mass of the reactants, and energy is produced.

12. Repeated fission reactions, or a ____________________, takes place when neutrons produced by earlier fission reactions cause the fission of other nuclei.

In your textbook, read about nuclear reactors.

Circle the letter of the choice that best completes each statement.

13. Most of the neutrons released by the fission of $^{235}_{92}U$ atoms are _____.

a. moving at slow speeds

b. slow neutrons

c. in pairs

d. moving at high speeds

14. When a $^{238}_{92}U$ nucleus absorbs a fast neutron, it _____.

a. becomes a new isotope

b. fuses with another atom

c. starts moving at a high speed

d. undergoes fission

15. Slow neutrons are more likely to be absorbed by _____ and to make a chain reaction possible.

a. $^{235}_{92}U$

b. $^{239}_{94}Pu$

c. $^{238}_{92}U$

d. $^{92}_{36}Kr$

16. When Fermi produced the first controlled chain reaction on December 2, 1942, he used _____ as a moderator.

a. hydrogen

b. water

c. heavy water

d. graphite

17. Enrichment increases _____ in the fuel for nuclear reactors.

a. the amount of gold

b. the number of fissionable nuclei

c. the number of fusionable nuclei

d. the number of slow neutrons

Name ______________________

31 Study Guide

In your textbook, read about the nuclear reactors used in the United States.

Refer to the drawing to answer the following questions.

Write labels for the drawing.

18. ______________________

19. ______________________

20. ______________________

21. ______________________

22. ______________________

23. ______________________

24. ______________________

25. ______________________

26. ______________________

27. ______________________

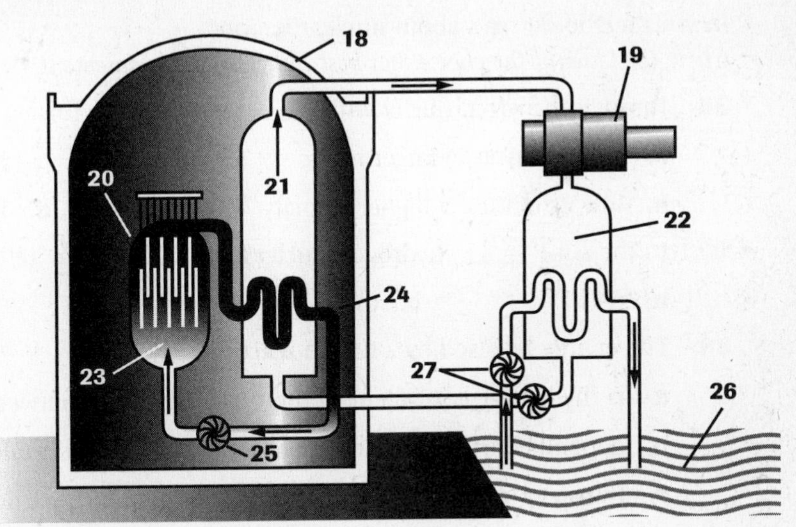

28. What type of reactor is shown in the drawing?

____

____

29. What purposes does water serve in this reactor?

____

30. Describe how the reactor produces electric energy.

____

____

____

31. Why do fuel rods glow blue when they hit water? What is this effect called?

____

____

____

In your textbook, read about the products of fission reactions.

For each of the statements below, write true *or rewrite the italicized part to make the statement true.*

32. ______________________ Old rods that can no longer be used in a reactor *aren't* radioactive.

33. ______________________ Plutonium could be removed from radioactive waste and *recycled* to fuel other reactors.

34. ______________________ The world's supply of uranium is *unlimited.*

Name ____________________

31 Study Guide

35. ____________________ From some reactors, *more fuel* can be recovered than was originally present.

In your textbook, read about nuclear fusion.

Circle the letter of the choice that best completes each statement.

36. In nuclear fusion, nuclei with _____ combine to form a nucleus with _____.
    a. small masses, a larger mass
    b. low velocities, a higher velocity
    c. large masses, a larger mass
    d. low energy, a higher energy
37. In the sun, _____ hydrogen nuclei fuse to form one helium nucleus.
    a. two
    b. three
    c. four
    d. six
38. The energy released by a fusion reaction _____.
    a. is the energy equivalent of the mass difference between the products and the reactants
    b. depends on the temperature at which the reaction takes place
    c. transfers to the potential energy of the resultant particles
    d. is very small compared to other types of reactions
39. The reaction $^{1}_{1}H$ + _____ $\rightarrow ^{3}_{2}He + \gamma$ is one possible step in the proton-proton chain of fusion in the sun.
    a. $^{1}_{1}H$
    b. $^{2}_{1}H$
    c. $^{2}_{1}He$
    d. $^{4}_{2}He$
40. In a thermonuclear bomb, the high temperature necessary to produce the fusion reaction is provided by _____.
    a. dynamite
    b. an atomic bomb
    c. a conventional bomb
    d. electricity

In your textbook, read about controlled fusion.

Answer the following questions, using complete sentences.

41. If energy could be produced economically by the fusion of deuterium and tritium, what would be two benefits?

____

____

42. What are two problems that prevent fusion reactions confined with magnetic fields from being used as an energy source?

____

____

43. Explain what happens during inertial confinement fusion.

____

____

____